D0874048

RAPTOR

RAPTOR

A JOURNEY THROUGH BIRDS

With a New Preface

JAMES MACDONALD
LOCKHART

THE UNIVERSITY OF CHICAGO PRESS
CHICAGO AND LONDON

The University of Chicago Press, Chicago 60637
The University of Chicago Press, Ltd., London

Published 2017
Printed in the United States of America

26 25 24 23 22 21 20 19 18 17   1 2 3 4 5

ISBN-13: 978-0-226-47058-0 (cloth)
ISBN-13: 978-0-226-47061-0 (e-book)

DOI: 10.7208/chicago/9780226470610.001.0001

LIBRARY OF CONGRESS CATALOGING-IN-PUBLICATION DATA

Names: Lockhart, James Macdonald, 1975– author.
Title: Raptor : a journey through birds : with a new preface / James Macdonald Lockhart.
Description: Chicago ; London : The University of Chicago Press, 2017. | Includes bibliographical references.
Identifiers: LCCN 2016043310 | ISBN 9780226470580 (cloth : alk. paper) | ISBN 9780226470610 (e-book)
Subjects: LCSH: Birds of prey—Great Britain. | Bird watching—Great Britain—Anecdotes. | Great Britain—Description and travel.
Classification: LCC QL677.78 .L632 2017 | DDC 598.072/3441—dc23
LC record available at https://lccn.loc.gov/2016043310

♾ This paper meets the requirements of ANSI/NISO Z39.48–1992 (Permanence of Paper).

Contents

	Preface	ix
I	*Hen Harrier*	1
II	*Merlin*	29
III	*Golden Eagle*	61
IV	*Osprey*	85
V	*Sea Eagle*	101
VI	*Goshawk*	133
VII	*Kestrel*	155
VIII	*Montagu's Harrier*	175
IX	*Peregrine Falcon*	195
X	*Red Kite*	211
XI	*Marsh Harrier*	233
XII	*Honey Buzzard*	253
XIII	*Hobby*	273
XIV	*Buzzard*	299
XV	*Sparrowhawk*	321
	Bibliography	357
	Acknowledgements	375

By the term Raptores may be designated an order of birds, the predatory habits of which have obtained for them a renown exceeding that of any other tribe ...

WILLIAM MACGILLIVRAY,
A History of British Birds, Volume III

Preface

One day in October 1830 the great American bird artist John James Audubon knocked on the door of a house in Edinburgh's New Town. Audubon was forty-five, tall, slightly stooped, long dark brown hair smoothed back above his forehead. Wearing a wolf-skin coat, walking quickly, he struck an odd sight on the streets of Scotland's capital. But Edinburgh had taken a shine to this exotic American who looked, dressed and spoke (his accent a curious blend of French and American) every bit the frontiersman.

This was Audubon's third visit to Edinburgh in the space of four years. On his first visit in 1826 his paintings had been exhibited in a room of their own at the prestigious Royal Institution. The exhibition was a huge success, *The Scotsman* newspaper described the paintings as 'all beyond praise', one picture was even stolen from the exhibition (later returned, found rolled-up on Audubon's doorstep). Before long everyone wanted to meet the great American wildlife artist. Coaches were sent to whisk Audubon off to dine with aristocracy. Sir Walter Scott (Audubon's hero) requested an audience with him. In January 1827 – a social and professional pinnacle – Audubon was made an honorary member of the Royal Scottish Academy. Uncomfortable in high society, many an evening Audubon found himself blinking with uncertainty in a chandeliered room in one of Edinburgh's great houses, missing his family in Louisiana, longing for letters from his wife Lucy. When her letters did finally reach him, Audubon wrote in his journal, 'How I read them! Perhaps never in my life were letters so well welcome, and they were such sweet letters...'

Though often homesick, it was his home country which had forced Audubon to travel to Britain in the first place. Audubon's ambition was to publish a definitive, illustrated guide to every species of bird in America. But his paintings had met with a frosty reception in Philadelphia and New York and the prospect of securing an American publisher for *The Birds of America* seemed remote. A friend encouraged Audubon to look instead for a publisher, as well as subscribers to help finance the publication, on the other side of the Atlantic. So in May 1826 Audubon set sail for England.

The reception to Audubon's paintings in Britain could not have been more different than it had been in the US. Soon after docking at Liverpool, Audubon exhibited his paintings at the Liverpool Royal Institution where, on the first day, over four hundred people came to see his work. In Edinburgh Audubon showed his paintings to the engraver William Home Lizars. Lizars was bowled over by the quality of Audubon's work, exclaiming, 'My God, I never saw anything like this before!' The meeting with Lizars in October 1826 was a turning point: Lizars agreed to engrave Audubon's paintings for him and early in 1827 the first volume of prints for *The*

Birds of America was published. However one crucial element remained missing from the project: as yet there was no text to accompany the illustrations of the birds. Audubon didn't feel he was up to the task of writing the text on his own. He needed to find and employ an ornithologist capable of assisting him in writing accurate descriptions of the birds, someone to correct his manuscripts and wobbly English (Audubon's first language was French). The first person Audubon approached to assist him, the English naturalist William Swainson, knocked him back, or rather asked too high a fee and also demanded that he receive joint accreditation on the title page. This wasn't what Audubon had in mind at all; he hadn't spent the best part of twenty years working on the book to 'transfer my fame to your pages & to your reputation...' as he put it in his reply to Swainson's offer. Then a godsend: a friend of Audubon's in Edinburgh, James Wilson, suggested Audubon look up William MacGillivray, a young naturalist beginning to make a name for himself in Edinburgh's scientific community. Furnished with a card with the address of Mr W. MacGillivray, Audubon wrote in his journal:

> ... and away to Mr. MacGillivray I went. He had long known of me as a naturalist. I made known my business, and a bargain was soon struck. He agreed to assist me, and correct my manuscripts for two guineas per sheet of sixteen pages and I that day began to write the first volume.

Audubon and MacGillivray would go on to work together on and off for the next nine years. The pair became close friends, writing together, walking together, shooting birds together. MacGillivray named one of his sons after Audubon; Audubon named two species of birds after MacGillivray, *MacGillivray's shore finch* and *MacGillivray's warbler*. In the gaps between writing Audubon travelled back and forth between America (to procure more bird specimens) and Britain (to write and garner subscriptions for the book). After each field trip to the United States, Audubon returned to Edinburgh and sought out MacGillivray to resume their writing. They worked intensely, Audubon starting work early, MacGillivray joining him later in the morning and then working late into the night.

The collaboration between Audubon and MacGillivray was, as I describe it in *Raptor*, like the coming together of an ornithological dream team. It was a collaboration which would produce one of the great founding works of American ornithology, the five volume *Ornithological Biography*, the accompanying text to Audubon's illustrated *The Birds of America*. MacGillivray's contribution to the *Ornithological Biography*, though largely forgotten now, was hugely significant. MacGillivray, not especially concerned with accreditation, did not receive any. Though subsequently others have acknowledged MacGillivray's crucial role, the American ornithologist Elliot Coues writing that MacGillivray 'supplied what was necessary to make his [Audubon's] work a contribution to science as well as to art'.

I've spent the last few years searching for and writing about all fifteen of the diurnal birds of prey which breed in the British Isles. I travelled to the Orkney islands in the far north of Scotland to study hen harriers, to a remote valley in Wales in search of red kites, to the Thames estuary to the east of London in pursuit of marsh harriers, and many other habitats, from mountains to motorways, in between. My book *Raptor* is the culmination of these journeys. It is a book about the birds and also a book about the places I went to search for the birds in. A common observation of Britain is that it's a small overcrowded island. Small compared to the great landmass of the United States, of course, but within this small island there is an extraordinary variety of landscapes, from the vast peat bogs found in the far north of Scotland to the deep wooded valleys of South West England. My book travels deeply into these places, into their history, their stories – and their birds.

Whilst I made these expeditions alone I don't feel I could have written this book without William MacGillivray. I have structured the book around an extraordinary eight hundred mile walk MacGillivray made from Aberdeen to London in September and October 1819, loosely mapping MacGillivray's journey against my own journey from the north of Scotland to the south of England. At various points in the book I pick MacGillivray up on his walk to explore his life and work and to think about his legacy. I came to rely greatly on MacGillivray. Whenever I found it difficult to find the birds (which was most of the time) or write about them accurately (which felt like all the time), I turned to MacGillivray, leaned on him, for support and guidance. If Audubon's meeting with MacGillivray in Edinburgh in October 1830 was a godsend then my own meeting with MacGillivray – encountering him through his numerous books, journals, his paintings of birds, the plant specimens he collected and carefully labelled nearly two hundred years ago – has equally been an inspiration to me, to this book.

History has not been as kind to MacGillivray as it has to Audubon. Today MacGillivray is a largely forgotten figure. Even his gravestone in Edinburgh was recently desecrated when the brass plaque of a golden eagle (modelled on one of MacGillivray's own paintings) was hacked off and stolen. By contract Audubon is renowned the world over and his paintings sell for vast sums (an original copy of *The Birds of America* was sold at auction in London in 2010 for $11.5 million). One of my hopes for my book *Raptor* is that it might contribute something towards redressing the balance, at the very least that it might help bring MacGillivray's name and work back into view.

One of the two American species of birds Audubon named after MacGillivray, *MacGillivray's shore finch* was later renamed the *seaside sparrow*. But the *MacGillivray's warbler* still bears his name. I doubt many people today will make the association between *MacGillivray's warbler* and the great nineteenth century Scottish naturalist. Though I for one am glad MacGillivray at least has a presence in the United States through the bird's name. I hope that the publication of my book *Raptor* in the United States will enhance MacGillivray's presence there. Through his collaboration with Audubon MacGillivray runs like an ornithological bridge across the Atlantic. I hope readers in the Unites States will cross that bridge and journey with me – and MacGillivray – through the birds of prey of the British Isles.

I

Hen Harrier

Orkney

It begins where the road ends beside a farm. Empty sacking, silage breath, the car parked amongst oily puddles. The fields are bright after rain. Inside one puddle, a white plastic feed sack, crumpled, like a drowned moon. Then feet up on the car's rear bumper, boots loosened and threaded, backpacks tightened. Wanting to rain: a sheen of rain, like the thought of rain, has settled on the car and made it gleam. When I bend to tie my boots I notice tiny beads of water quivering like mercury on the waxed leather. Eric is with me, who knows this valley intimately, who knows where the kestrel has its nest above the burn and where the short-eared owls hide their young amongst the heather. We leave the farm and start to walk along the track towards the swell of the moor.

Closer the fields look greasy and soft. The track begins to leak away from under us and soon the bog has smothered it completely. We are amongst peat hags and pools of amber water. Marsh orchids glow mauve and pink amongst the dark reed grass. The sky is heavy with geese: greylags, with their snowshoe gait, long thick necks snorkelling the heather. You do not think they could get airborne; they run across the moor beating at the air, nothing like a bird. And with a heave they are up, calling with the rigmarole of it all, stacking themselves in columns of three or four. They fly low over the moor, circling above us as if in a holding pattern. When a column of geese breaks the horizon it looks like a dust devil has spun up from the ground to whirl slowly down the valley towards us.

Late May on a hillside in Orkney; nowhere I would rather be. It is a place running with birds. Curlews with their rippling song and long delicate bills and the young short-eared owls keeking from their hideout in the heather. And all that heft and noise of goose. When the greylags leave, shepherding their young down off the moor, following the burns to the lowland lochs and brackish lagoons, then, surely, undetectably, the moor must inflate a little, breathing out after all that weight of goose has gone.

We find a path that cuts through a bank of deep heather. It leads up onto the moor and the horizon lifts. I can see the hills of Hoy with their wind-raked slopes of scree and the sea below with its waves like the patterns of the scree. This morning the sea is a livery dark, creased with white lines that map the movement of the swell. It looks as if the sea is full of cracks, splinters of ice.

Wherever you turn on Orkney the sea is at your back, linking the islands with its junctions of light. It is not enough that the islands are already so scattered. The sea is always gnawing at them, looking for avenues to open up, fractures in the rock to prise apart. The sea up here has myriad ways to breach the land. It showers the western cliffs with its salty mists and peoples the thin soils with its kin: creeping willow, eyebright, sea thrift, sea plantain, all plants that love the sea's breath on them.

It is a trickster sea that comes ashore with subterfuge. Orkney children once made imaginary farms with scallop shells for sheep, gaper shells for pigs, as if the sea, like a toymaker, had carved each shell and left it on the shore, waiting for a passing child. And at night selchies dock in the deep geos and patter ashore in their wet skins to slip amongst the dozing kye.

I arrived on Orkney in the dregs of a May gale. Low pressure swelling in from the Atlantic, hurting buildings and trees in their new growth, ransacking birds' nests. Rushing across Scotland and speeding up over Orkney as if the gale had hit a patch of ice. The hardest thing of all up here, I'd heard, was learning to endure the wind, worse than the long winter darkness. It is a fidgety wind, rarely still, boisterous, folding sheds and hen houses, raking the islands' lochs into inland seas.

This morning the wind still has a sinewy strength. Lapwings are lifted off the fields like flakes of ash. Eric is telling me about the valley. He is alert to the slightest wren-flick through the heather, seeing the birds before they arrive. When he speaks the wind gashes at his words, gets inside them. The hill is shaking with wind. We pass the kestrel's nest and Eric points out a clump of rushes on the hillside where a hen harrier is sitting on her eggs. She is invisible on her bed of rush and ling. Her eggs are pale white, polished, stained with the colours of the nest material. If you could see like a hawk you would notice her bright yellow eyes, framed by a white eyebrow on a flat owl-like face. The face made rounder by the thick neck ruff that flickers bronze and almond-white like the ring around a planet.

Then Eric is leaving and I don't feel I have thanked him enough. I watch him descend the moor, walking quickly along the gleaming track towards the farm and the car like a skiff bobbing amongst the lit puddles. After he has gone there is a sudden rush of rain. But the wind is so strong it seems to hold the rain up, stops it reaching the ground, flinging the shower away, crashing it into the upper slopes of the hill. Only my face and hair are briefly wet; the rest of me stays dry, as if I'd poked my head into a cloud. I have never seen rain behave this way. Months later I came across a list of beautiful old Orkney dialect words for different types of rain and wondered which described the behaviour of that wind-blown shower: *driv, rug, murr, muggerafeu, hagger, dagg, rav, hellyiefer* ... A *rug*, perhaps, meaning 'a strong pull', rain that was being pulled, yanked away by the wind.

I had not thanked Eric nearly enough. For a walk like that will have its legacies, store itself in you like a muscle's memory. Walking up through layers of birds, Eric explaining the narrative of the moor, where last year's merlins hid their nest, where cattle had punctuated the dyke and damaged the delicate hillside. Till we reached a fold in the hills, the 'nesting station', where the hen harriers had congregated their nests, and we could go no further.

I know of other walks, like the one with Eric that morning, where their legacy is precious and defining, walks born out of that experience of guiding or being guided. My great-grandfather, Seton Gordon, in the early summer of 1906, when he was only twenty, walked into the Grampian Mountains with his boyhood hero, the naturalist Richard Kearton. That walk began with Gordon telegramming Kearton with the news he had found a ptarmigan's nest, one of the few birds, Gordon knew, that Kearton had never photographed. Kearton packed hastily and rushed to catch the next train to Scotland. Early June, travelling north through strata of light; a 600-mile journey from Surrey to Aberdeenshire, where Gordon met Kearton off the train at Ballater. They decide to climb the mountain at night to avoid the heat of the day, setting off in the dusk, the smell of pine and birch all around them. Kearton is lame (he was left permanently lame after a childhood accident) and has to walk slowly, stopping often to rest. They toil up the mountain through the thin June dark, Kearton bent like a hunchback under the huge camera he is carrying (the heaviest Gordon has ever seen). At 1.45 a.m. a redstart's song spurs them on.

They reach the snowfield beside the ptarmigan's nest at 6 a.m. The sun is up and bright, the short grass sparkles. Kearton assembles his camera on its tripod and begins, cautiously, to crawl towards the sitting bird. She is a close sitter, Gordon reassures him, and if he stalks her very slowly she should sit tight. The next few moments are so precarious: Kearton exposes a number of plates and after each exposure he edges a little closer towards the ptarmigan. He stops when he is just nine feet away. He can hear his heart thumping in his chest. One last exposure, that's it! He is close enough to see the bird breathing and the dew pearled across her back.

Seventy years later, the year before he died, Gordon was still able to recall that walk, writing about it in an article for *Country Life* magazine. The details of that day still fresh and resonant: the brightness of the sun that morning, the dew along the ptarmigan's back, the cost of the telegram he sent to Kearton (sixpence).

Richard Kearton's photographs of birds, taken at the end of the nineteenth century (many with his brother, Cherry Kearton), were to my great-grandfather what Gordon's own photographs of birds are to me, jewels of inspiration. I grew up surrounded by Gordon's black and white photographs of birds: golden eagles, greenshanks, gannets, dotterels ... peering down at me from their teak frames. I liked to take the frames off the wall, wipe the dust from the glass, then turn the pictures over to read the captions Gordon had written on the back of each print:

- Female eagle 'parasoling' eaglets. The eaglet is invisible on the other side of the bird.
- The golden eagle brings a heather branch to the eyrie.

Whenever I have moved house the first pictures I hang on the new walls are two small photographs Gordon took, one of a jackdaw pair, black and pewter, the other of a hooded crow in its sleeveless silver waistcoat. Under a cupboard I keep a great cache of Gordon's photographs in an ancient marble-patterned canvas folder. You have to untie three string bows to open the folder, and every time I do so a fragment of the canvas frays and disintegrates. My young children like to open the folder with me, and the process of going through the photographs with them – identifying the birds and mammals – has become a lovely ritual. The photographs are beautiful. I am still amazed that anyone could get so close to a wild bird as Gordon did, and photograph it in such exquisite detail. In one photograph, taken in 1922, a golden eagle lands on its nest with a grouse in its talons as a cloud of flies spumes out from an old carcass on the eyrie, as if the landing eagle has triggered an explosion. There is a stunning photograph he took of a pair of greenshanks just at the moment the birds change over incubation duties at their nest. One bird steps over the nest, ready to settle, as its mate pulls itself off the clutch of four eggs. The timing of the photograph, to capture the precise moment of the changeover, is extraordinary. The patterning on the eggs matches the patterns down the greenshank's breast as if one has imprinted – stained – the other.

Along with Richard Kearton and another wildlife photographer, R. B. Lodge (both important influences on Gordon), Gordon was in the vanguard of early bird photography in this country. Cycling around Deeside in the first years of the twentieth century with his half-plate Thornton Pickard Ruby camera with Dallmeyer lens, Gordon took many exceptional photographs of birds and the wider fauna of the region. Upland species were his speciality: snow bunting, curlew, red-throated diver, ptarmigan ... Many of these photographs he published in the books he wrote. Twenty-seven books in all, the bulk of them about the wildlife and landscapes of the Highlands and Islands. His books take up a wall of shelving in my house; greens and browns and pale silver spines, embossed with gold lettering: *Birds of the Loch and Mountain*; *The Charm of the Hills*; *In Search of Northern Birds*; *Afoot in the Hebrides*; *Wanderings of a Naturalist*; *The Cairngorm Hills of Scotland*; *Amid Snowy Wastes*; *Highways and Byways in the West Highlands* ... I love their Edwardian-sounding titles, the earthy colours of their spines, the smell and feel of the books' thick-cut paper with its ragged, crenulated edges.

Of all the birds that Gordon photographed and studied, the golden eagle was the one he came to know the best. Eagles were his abiding love, his expertise. He published two monographs on them: *Days with the Golden Eagle* in 1927 and *The Golden Eagle: King of Birds* in 1955. Both these studies, particularly the latter, went on to influence and inspire a subsequent generation of naturalists and raptor ornithologists. I often come across warm references to Gordon's golden eagle studies in the forewords and acknowledgements of contemporary works of ornithology. His two golden eagle books were also a huge influence on me. I read them many times when I was a teenager. They set this book gestating.

And then there's my own version of Gordon's walk with Kearton. A family holiday on the Isle of Lewis. I am fourteen or fifteen. In the photos from that week we are sitting on the island's vast, empty beaches, our hair washed out at right angles by the wind. Family picnics: brushing the sand off sandwiches, a lime-green thermos of tomato soup. In the background, sand dunes, a squall inside the marram grass, a blur of gannet flying west.

One day out walking on the moors I discovered something momentous. I had been following a river into the hills and as I came up over the watershed I noticed a small loch lying in a shallow dent of the moor. Wherever there is a depression in the land on Lewis, water gathers. It patterns the island intricately, beautifully. From the air much of the land looks tenuous, as if it is breaking apart, a network of walkways floating on black water. The loch I came across – a lochan – a small pool of peat-dark water, like a sunspot against the purple moor. In its centre, rowing round and round, beating the lochan's bounds, a red-throated diver with its chick. I was mesmerised. They are beautiful, rare birds that I had read about but never seen. I sat down in a bank of heather above the loch and watched the bird's sleek outline, the faint blush of her throat. Always the sense in her streamlined shape that, sitting on the water, she was not quite in her element, that once she dived, like an otter, the water would transform her. But I had to get back – needed to get back – and tell someone. I raced over the moor and gasped out my discovery to nodding, distracted faces.

Except for Mum. She was interested. She wanted to hear more about what I had found, got me to show her in a bird book what the diver looked like, listened when I explained how its eerie, otherworldly call was supposed to forewarn of rain. And the next morning it rained but Mum and I walked across the moor to the lochan where I had found the divers. And so I became a guide, leading my mother up the river and over the rise in the moor, and in doing so felt something Gordon must have felt guiding Richard Kearton up the flank of the mountain through the night. Me walking far too fast in my exhilaration, almost running over the peat bog, Mum calling me to slow down. Reaching the lochan: the two of us sitting down in the heather, catching our breath. Mum asking if she could borrow my binoculars.

* * *

Where Eric left me became my home for the next two days. I walked up the valley in the early morning, looking for the print my weight had made in the heather the day before; not recognisably my shape, more like a shallow quaich scooped out of the heather as if snow had slept there and in the morning left and left behind its thaw-stain on the ground. The heather held me almost buoyant in its thickness. I settled down in it like a hare crouched in its form, felt the wind running over my back.

Rummaging through Orkney's deceased dialects I borrowed a handful of words besides those I had found to describe the different kinds of rain. Loaned words to help me navigate the land. I liked the ways, like a detailed map, they attended to the specifics, to the margins of the landscape around me. *Cowe*: a stalk of heather, which the wind swept like a windscreen wiper across my view of the moor; *burra*: hard grass found in moory soil; *gayro*: the sward on a hillside where the heather has been exterminated by water. And this word – a lovely gift – which described perfectly my form in the heather, *beul*: a place to lie down or rest.

To reach my *beul* on the moor I had to pass through different zones of birds. Each species seemed to occupy its own layer of the valley as if it adhered to an underlying geology of the place. Oystercatchers over a layer of marl, curlews spread across a bed of sandstone. All of these territories seeping, blurring into each other along fault lines in the moor. For a long time I struggled to get a hold of the birds. There was so much movement amongst them, so many birds to keep track of. I began to draw Venn diagrams of the birds' locations and their movements across the moor till the pages of my note-book were filled with overlapping water rings. Gradually I came to see this small patch of moorland as the meeting point of several territories. There were four species of raptor alone breeding in the heather around me – kestrel, hen harrier, merlin and short-eared owl – and these birds' interaction with each other, and with the large population of breeding wader birds, the curlews in particular, held me captivated over the two days I was there.

Trespassing, ghosting through all these territories like blown fragments of white silk, were the short-eared owls. The focus of their territory was a shifting area that moved around the location of the young owls as they dispersed through the deep heather. Sometimes I passed close by the owlets, hidden from me, calling loudly to their parents for food. Their high-pitched *keek* drilled through me as if I was passing through a scanner. Then the adult birds appeared beside me. They arrived suddenly, quietly, like shapes congealed out of mist. They flew very close, hovering just above my head. Pale white underwings marked with black crease lines as if in places the wings were beginning to thaw. Their faces a deep snow-cloud grey. Black eye-bands, like masks, which set their bright yellow eyes deep in their cupped faces. Always, as the owls hung above me, a sense of being watched and of their gaze penetrating through me. Beautiful in their buoyancy. Wing-tippers, possessed of twisting grace, flickering low over the moor like giant ghost moths.

I never expected to see so much so soon on this journey. I had set off fully expecting to be frustrated, to not find many of the birds I was looking for. But Orkney gave me so much time with its birds of prey, spoilt me utterly in that respect, so that every stage along my journey since has harked back to the time I spent on the islands.

There was an old charm used on Orkney which was supposed to heal deep cuts to the skin. To initiate the cure you had to send the name of the wounded person to the local enchanter or witch-doctor (the Old Norse term was *vǫlva*, meaning a wand carrier). The *vǫlva* would then add some new words to the patient's name, as if they were sprinkling ingredients into a recipe. This word-concoction was then chanted repeatedly until, through some telepathic alchemy, the wound was healed. The charm worked even though the *vǫlva* performing it might be several miles away from their patient. After I left Orkney – and was many hundreds of miles away from the islands – I often wanted to send word back to the place, hoping the islands could perform their charm on me once more and help me find the birds I was searching for, just as they had gifted them to me while I was there.

How does a hen harrier live? It swims over the land as a storm petrel hugs the surface of the sea. It flies so low that sometimes it seems to be stirring the grass, its long legs trailing through the heather like a keel. A slow, tacking flight: float then flap. Then a twisting pirouette and it has swung onto a different tack, following another seam through the moor as if it is tracking a scent. It is like watching a disembodied spirit searching for its host, like a spirit twisting and drifting in the breeze. The harrier rides the swell of the moor, clinging to the contours for cover, creeping up on its prey and surprising it with a sudden burst of speed. The moment before it swoops, the harrier stalls, adjusts its angle a fraction – kestrel-like – then drops suddenly to the ground, talons reaching, seizing through the deep grass.

Up on the moor a cold wind has found me, poking me for signs of life. It sets me shivering in my bed in the heather. A male hen harrier is taking his time, twirling a thread over the hill. He is the colour of smoke and lavender, glaucous. He has not killed. The female is up from her nest agitating him, calling repeatedly, pushing him away from the nest site, chivvying him to go and hunt. The male slips away down the valley towards the sea. Later in the morning, when he returns, flapping heavily up towards the moor, he seems brighter, sea-gleamed, as if the sea has dressed him in its brightness.

I stay with the female harrier expecting her to quickly settle on the nest. Instead, what happens next is startling, unexpected, one of those remarkable encounters that Orkney shared with me. She began to gain height, drawing herself up to just above the horizon. Then a pause as she floated there as if in a new-found buoyancy, as if on the apex of a thermal. Perhaps the thermal had a vacuum like a lift shaft running through its centre. Whatever it was, a route through the air opened up to her like a narrow channel – a lead – opening through pack ice. For she saw it, drew her wing tips back behind her, and plunged down through the opening in a beautiful corkscrew dive. At the last moment, before she crashed into the heather, swooping up again, rising above the moor, clearing the horizon, then leaning and tipping over into another dive. She was dancing! A mesmerising skydance, repeated over and over again, making a pattern in the air like the peaks and troughs of a frantic graph. I paused from watching her only to wipe the condensation from the lenses of my binoculars. Then I immediately fixed back on to her display. Was she signalling to the male, or simply flexing her flight muscles after a long stint on the nest? For twenty minutes she scored the air, and held me there.

The hardest thing to do was to leave the moor in the evening. I walked down through the restless geese and kept turning round in search of one last glimpse of the hen harriers. Sometimes I would follow a male harrier down off the peat moss and watch him quartering the marshy borderland between the fields and the moor. Much of Orkney's moorland has been reclaimed by agriculture. Only the difficult land remains as moor, but even this is sometimes coveted: there are hills on Orkney with strips of neat green fields dissecting the moor. In places it looks like the hills have been scalped. But in the borderlands between the moor and the fields there seemed to be a tussle between the two spheres so that it was unclear which was reclaiming the other. What had been tidied from the fields seemed to have been swept to these edges. And hen harriers thrived in these places, hunting voles along these tangled, unkempt margins.

I met hen harriers in unlikely places on the island. Far away from the moors, around the backs of houses, through the engine-ticking quiet of a farmyard. The birds seemed to slip easily between spheres, between the hills and the farmland below. In an evening field behind a white hotel, I watched a great commotion of oystercatchers and curlews dive-bombing a female harrier. She was slow and huge amongst the shrill wading birds, like some wandering beast come down out of the hills to forage.

Driving across the hill road to Harray in the early evening, I came down off the moor towards a farm and there was a male hen harrier swimming in a pool of wind. I stopped the car and watched him swirl through the farmyard, low over a hedge and into a back garden. He drifted up over a washing line and, for a moment, seemed to join the garments there, a blue-grey shirt flapping in the wind.

This morning the wind has shed some of its weight. The curlew's song has more reach. A male harrier is coming in from the west, lucent against the heather. He is flying more quickly than usual, keeping a straight line, heading for the nest that sits in the lap of a hill amongst thick tussocks of moor grass. Now the female is up and rising to greet him, rushing towards the male. She is so much larger than him, her colours so markedly different, her tawny browns set against his smoking greyness. For centuries the male and female hen harrier were thought to be a different species, and this morning she might have been a larger hawk about to set upon the smaller male. But, at the last moment, she twists onto her back beneath the male and their talons almost brush. The male releases something from his feet and she seizes and catches it in mid-air. All of this happens so quickly and the movement is astonishing for its speed and precision. I cannot make out what it is the male has passed to her but the female has flipped instantly upright again and is rising towards the male once more. Again, at the last moment, she twists onto her back but this time nothing is passed between them. I'm puzzled why she repeats the manoeuvre like this. Perhaps the male still has something in his talons? But I am lucky to witness it again, it is like an unexpected echo and gives me the chance to replay the whole extraordinary exchange. I stay with the female and watch her drop into the heather, where she begins to feed.

Why do hen harriers make this beautiful, acrobatic food pass? When she is incubating, then brooding and guarding the young at the nest, the female is dependent on the male to provide food for her and the chicks. Later, when the chicks are closer to leaving the nest, she will resume hunting. But until then the male must work overtime to provide for the brood and his mate. Polygamous males, common on Orkney, are required to ratchet up their hunting, providing for two, sometimes three, separate nests (the record on Orkney is seven). Nesting in tall dense vegetation on the ground, the food pass is the most efficient way of securing the exchange of prey whilst also distracting from the precise location of the nest, the pass often taking place some distance from the nest itself, which helps to avoid drawing its attention to predators. I wonder if the female ever drops the pass from the male. The hen harrier is so supremely agile, their long legs have such reach, it seems like the male could lob the most awkward pass at his mate and she would still pluck the prey out of the air with all the time to spare.

On Orkney, the Orkney vole (twice the size of the field vole found on the British mainland) is a crucial prey species for the hen harrier. The relatively high numbers of hen harriers on the islands (on those islands in the archipelago that have voles) is attributed to the abundance and stability of the vole population. Many hen harriers overwinter on Orkney and voles are the principal reason these birds are not forced to migrate further south. In addition to voles, young curlews and starlings are frequently taken on the islands, as well as meadow pipits, skylarks and lapwings. Rabbits, too, are predated by the larger female harrier. Where voles are absent, hen harriers are able to breed as long as there is an abundant supply of passerine birds. But on Orkney, and over much of the hen harrier's range, avian prey is, on the whole, secondary in importance and preference to voles and other small mammals, so much so that a scarcity of voles can impede the hen harrier's breeding success. In Gaelic the hen harrier's name is *Clamhan luch* (the mouse-hawk).

Midday and the moor is quiet, the slackest time of the day. The harriers are sitting tight. At Lammas time, under a full moon, people used to go up to the moors on Orkney to cut the stems of rushes to use as wicks for fish-oil lamps. And they went to the moors to gather the wiry cowberry stems to twist into ropes. The moors were a busier landscape than they are today, more interacted with. There was a steady trafficking of peats from off the moors: all the different grades of Orkney peat, peats that smelt of sulphur when they burnt, heavy *yarpha* peats with the moss and heather still on them, peats that burnt too quickly and left behind a bright creamy ash in the hearth. Geese (an important part of the economy on Orkney right up till the mid-nineteenth century) were brought down from the moors when they became broody and taken into people's homes. Most houses on the islands were designed to accommodate the geese, with a recess cut into the wall beside the hearth where they were lodged to incubate their eggs in the warmth.

I get up to stretch my legs and go for a wander through the network of peat hags. I liked the notion of geese being 'let in' to people's homes, of the architectural twist made to houses to accommodate the birds. In the Hebrides, if someone had an especially lucky day, it was said that they must have seen the *Clamhan luch*. In Devon the hen harrier was known as the *Furze hawk*; in Caithness the *Flapper*; *Hebog llwydlas* (the blue-grey hawk) in Wales;

Saint Julian's bird in South Wales; in the English Midlands the *Blue hawk* ... We ought to let these birds back in. Today you can count the number of hen harriers nesting in England on your hand. And each year lop another finger off. They are not where they should be and their absence in the English uplands is shameful, a waste. A landscape devoid of hen harriers is an impoverished one. Hen harriers do predate red grouse – young grouse and weaker adult stock are the most vulnerable. But management of grouse moors and protection of hen harriers should not – does not need to – be incompatible. We need to let the harriers back in. Because a bird like this can change the way you see a landscape. Because (I promise you this) these birds will astonish you with their beauty. I wish others could see what I saw over Orkney, how the harriers made a ballet of the sky. I wish more people had the chance to see how the black wing tips of the male hen harrier are offset – made blacker – by his pearl-grey upper wings.

Some of the peat hags are so deeply cut into the moor it is like walking through a trench system. I can move across a flank of the hill without being seen. Except, of course, the short-eared owls have seen me. One of the adult birds has swum over to hover above my head. I can see the flickering gold and black patterns of its plumage, the gold like a dusting of pollen over its feathers. I sit down on the peat bank while the owl's shadow grazes over me. Earlier, I watched one of the owls and a male hen harrier hunting over the same patch of moorland. The owl seemed to hold something back. At least, the owl was not quite as fluid as the male harrier, who appeared to give himself over completely to the wind, like an appendage of the wind, sketching its currents and eddies, its tributaries of warm air.

Later: the moor is woken by a loud clapping noise. I peer over the trench and an owl is spiralling downwards, 'clapping' his wings together as he descends, signalling, displaying to his mate, bringing his great wings down beneath him as if he were crashing a pair of cymbals, whacking the air to show how much he owns it.

I follow the line of an old fence across the hill. Beneath one of the fence posts I find a small pile of harrier pellets. They look like chrysalids, parcels of hair wrapped around hints of bone as if something were forming in there. I can make out the tiny jawbone of a vole with a row of teeth along its edge, like a frayed clarinet reed.

The quiet ran on through the early afternoon. I hadn't expected this, that the moor could shush itself and doze in the day's thin warmth. How does a bird that was once seen as a harbinger of good fortune in the Hebrides become so reviled? Three hundred and fifty-one hen harriers are killed on two estates in Ayrshire between 1850 and 1854; one keeper on Skye kills thirty-two harriers in a single year in 1870; another, on the same island, accounts for twenty-five hen harriers in 1873. An article on Highland sports in *The Quarterly Review* of 1845 illustrates the attitudes of the day:

> Hawks of all sorts, from eagles to merlins destroy numbers [*of game*]. The worst of the family, and the most difficult to be destroyed is the hen harrier. Living wholly on birds of his own killing, he will come to no laid bait; and hunting in an open country, he is rarely approached near enough to be shot: skimming low, and quartering his ground like a well-trained pointer, he finds almost every bird, and with sure aim strikes down all he finds.

Though not so *difficult to be destroyed* as this article posits. The hen harrier, in fact (as Victorian game-book records testify), made an easy target: a large, slow-flying, ground-nesting bird with a tendency, amongst the females especially, to be fearless around humans when defending their nest. Female hen harriers are not unknown to dislodge hats, even scrape a person's scalp with their talons, should they venture too close to the nest.

A male harrier drifts along the horizon. He lands on a fence post and begins to preen. The fence follows the horizon and the harrier, perched there, is silhouetted against the backdrop of sky. He glimmers there. Then he drops, pirouettes, hesitates a few feet above the moor and lunges into the grass. It is a quick, purposeful drop, not like the half-hearted pounces I have seen. I know straight away he has killed. I can see him in the grass plucking, tearing at something. After this, he feeds for several minutes. Then he is up and carrying the prey in his talons, flying direct to the nest site. And there is the female rising, making straight towards him.

My last day on the islands. I decide, reluctantly, to leave the Orkney Mainland and its hen harriers and travel to the southernmost isle in the archipelago, South Ronaldsay. I have heard about a place on the island called 'The Tomb of the Eagles', a Neolithic chambered cairn overlooking the cliffs at Isbister. And, well … the tomb's name is enough to make me want to visit.

Late morning and I am walking out across the fields towards the cliffs at Isbister. The heath shimmers in the warm air. In the distance, a broken farmstead surfaces out of the heath like a whale. When I find the tomb I lie down beside the sea pink and begin to crawl along the narrow tunnel that leads into the tomb's interior. Inside, it is a nest of cool air. The stone walls are rent with algae sores, green and verdigris capillaries. I make my way to the far end of the tomb and duck into a side chamber; it has a moonscape floor, sandy, strewn with pebbles. I sit down inside the cell with my back against the rock.

In 1958, Ronnie Simison, a farmer from South Ronaldsay, was walking over his land looking for stone to quarry for use as fence posts. He walked along the sea cliffs at the eastern edge of his farm. Below him fulmars were nesting on sandstone ledges, a seal was berthing in Ham Geo, curlews moved amongst marsh orchids and eyebright. Perhaps it was the pink splash of sea thrift that caught his eye and drew him to the arrangement of stones the weather had recently exposed, a grassy mound peeled back to reveal a sneak of wall. A spade was fetched and Simison began to dig down beside the wall. As he dug the stones spilt things about his feet as if his spade had disturbed a crèche of voles. He picked each object up and laid it out beside him on the grass: a limestone knife; a stone axe head; a black bead, polished and shiny, that stared back at him like an eye. Then his spade had found an opening and darkness was spilling out of a doorway like oil. He fumbled in his pockets and pulled out a cigarette lighter, stretched his arm into the darkness and flicked the lighter's wheel. The bronze shapes he saw flickering back at him must have made him nearly drop the lighter. Certainly, the story goes, Simison ran the mile back across the swaying grassland to his home where, breathless and sweating, he picked up the telephone and called the police.

Inside the tomb I can hear a curlew trilling above the heath. I crawl back down along the tunnel and out into the bright sea glare. My trousers are covered in dust from sitting on the cell's floor and, as I walk along the cliffs, it looks like my legs are smoking as the breeze cleans the dust from my clothes.

The darkness Ronnie Simison's spade had cut loose from the mound that summer's evening was 5,000 years old. The mound was a Neolithic chambered cairn, and staring back at Simison when he sparked his lighter into the dark hole was a shelf of human skulls, resinous in the flickering light. There might have been a second or two when Simison mistook the bones' bronzed colours for a cache of treasure before he realised what they were, grabbed his spade, and ran.

When the tomb came to be excavated, amongst the human bones it was discovered that there were many bones and talons belonging to white-tailed sea eagles. In all, seventy sea-eagle talons were found and, in some instances, the birds' talons had been placed beside the bones of human individuals (one person had been buried with fifteen talons and the bones of two sea eagles). It is estimated that there were thirty-five skeletons of birds of prey in the tomb and of these two-thirds belonged to sea eagles.

The sea eagle was clearly a bird of totemic significance for the people living in that part of Orkney at that time. Presumably the bird performed some sort of funerary or shamanistic role for the community, perhaps in accompanying the dead on their journey to the afterlife, perhaps in assisting shamans in their magico-religious ceremonies. The importance of birds in shamanistic rituals is well known and there are archaeological examples from different cultures around the world of birds being involved in ceremonial and mortuary practices. In Alaska archaeologists unearthed a grave from a proto-Eskimo settlement at Ipiutak in which an adult and a child had been interred alongside, amongst other artefacts, the head of a loon (a species of diver). Strikingly, the diver's skull had lifelike artificial eyes (carved ivory for the white of the eye inlaid with jet for the black pupils) placed in its eye sockets. It's possible these ivory eyes served as a prophylactic to ward against evil (some human skulls from the settlement also contained artificial eyes). Equally, the eyes may have been placed in the diver in recognition of the belief amongst circumpolar peoples that the loon, a totemic bird for these cultures, was a bird with the power to both restore sight and also assist shamans with seeing into – and travelling through – different worlds.

In the museum a mile from the tomb some of the skulls have been given names: 'Jock Tamson', 'Granny', 'Charlie-Girl'. Beside the skulls there were pieces of pottery, fragments of bowl decorated by the imprint of human fingernails. The nails had scratched the wet clay and left a pattern like a wavy barcode around the bowl's rim. I picked some of the sea-eagle talons out of their case and held them in my palm, running my fingers over their blunted points. They were smooth to touch, like polished marble, their creamy colours flecked with rust.

The human bones, in contrast to the eagle and other animal bones in the cairn, were found to be in poor condition, noticeably bleached and weathered. This weathering suggests that the human dead were excarnated, given 'sky burials', their bodies exposed to the elements on raised platforms to be cleaned by natural decay and carrion feeders like the sea eagle. Besides the eagle bones, which were by far the most numerous, there were also bones of other carrion-feeding birds inside the tomb: two greater black-backed gulls, two rooks or crows and one raven. Once the excarnation process had been completed, the human skeletons – their bones scattered by carrion birds and bleached by the sun – would have been gathered up and interred inside the tomb.

That the sea eagles were involved in the excarnation of the human dead on Orkney is almost certain given that the bird is such a prodigious carrion feeder. Excarnation: the separation (of the soul) from the body at death, the opposite of incarnation, where the soul or spirit is clothed, embodied in flesh. Excarnation is not just a method for disposing the dead (to excarnate means to remove the flesh). It was also, for some societies, the process by which the spirit or soul could be released from the flesh. Tibetan Buddhists believed that the vultures summoned to a sky burial were spirits of the netherworld come to assist the soul on its journey to its next incarnation. In parts of the Western Highlands of Scotland it was unlucky to kill seagulls because it was believed the birds housed the souls of the dead. For what better, more natural place to rehome the soul – the restless, fidgety soul – than a bird, whose shape and movement, whose own restless flight, could be said to resemble the soul? Perhaps the Neolithic peoples of Orkney believed something similar, that when sea eagles, this great totemic bird, cleaned the bodies of their dead, the person's spirit, which after death still lived on inside the flesh, was taken in by the eagle. The spirit or soul transmigrated to the bird, lived on inside the bird. Human and eagle fusing – literally, ceremonially – each one inhabiting the other.

II

Merlin

The Flow Country

A man is walking south from Aberdeen to London. The night before he leaves his home in Aberdeen he dreams of birds, row upon row of birds perched in their glass cases in a locked museum. He walks along the museum's deserted corridors, his footsteps scurry ahead of him. He wants to slow the dream, to pause and study every specimen. When he looks at a bird he looks inside it, thinks about the mechanics of it, how it works, the map of its soul. But the sterility of the place makes him itchy and the dream begins to tumble into itself as it rushes towards its closing. The birds wake inside their cabinets and start to tap the glass with their beaks. The noise of their tapping: it is almost as if the birds are applauding him. Then some of the cabinets are cracking and the birds are prising, squeezing through the cracks. A mallard drake cuts itself on a shard of glass and he sees its blood beading black against the duck's emerald green. Then birds are pouring past him and the museum's roof is a dark cloud of birds. And he – William MacGillivray – is flickering awake.

It is 4 a.m., a September morning in 1819. William MacGillivray is twenty-three, fizzing, fidgety within himself. He writes: *I have no peace of mind*. He means: he is impatient of his own impatience. Often he is cramped by melancholy. In his journals he checks, frequently, the inventory of himself; always there are things missing and the gaps in his learning gnaw and grate. Travelling calms him, it gives him buoyancy, space to scrutinise his mind. And so he walks everywhere, thinks nothing of a journey of 100 miles on foot through the mountains. He recommends liniment of soap mixed with whisky to harden the soles of restless feet. His own feet are hard as gneiss, they never blister.

He is not unlike a merlin in the way he boils with energy. He once watched a merlin pursuing a lark relentlessly over every twist and turn. The pair flashed so close to him he could clearly see the male merlin's grey-blue dorsal plumage. The tiny falcon rushed after the lark, following it through farm steadings, between cornstacks, amongst the garden trees.

He lives his own life much like this, restless, obstinate, plunging headlong after everything. Not unlike a merlin, too, in the patterns of his wanderings, seasonal migrations. Leaving his home on the Isle of Harris to walk – at the beginning and end of every term – back and forth to university in Aberdeen, sleeping under brooms of heather, in caves above Loch Maree. Most of what he knows about the natural world – in botany, geology, ornithology – he has learnt from these walks. He can name all the plants that grow along the southern shore of Loch Ness. Often his walks digress into curiosity. He will follow a river to its source high in the mountains just to see what plants are growing there. Other times he eats up the distance, 40–50 miles in a day. If he stops moving for too long it's not that his mind begins to stiffen, rather that it trembles uncontrollably.

And now this long walk to London. Because: his mind is a wave of aftershocks and he is desperate – has been desperate for weeks – to be away from Aberdeen, to be out there on the cusp of things. In his house in the city he is tidying away his breakfast, crumbs from a barley cake have caught on his lip. Then a final check through the contents of his knapsack. He calls his knapsack *this machine*. It is made of thick oiled cloth and cost him six shillings and sixpence. Inside the *machine*: two travelling maps, one of Scotland, one of England; a small portfolio with a parcel of paper for drying plants; a few sheets of clean paper, stitched; a bottle of ink; four quills; the *Compendium Flora Britannica* ... He picks up the knapsack; its cloth is stiff with newness, like a frozen bat. For a while he tries to knead the stiffness out of the straps. His hands smell like a saddler's.

Five a.m. Outside in the street the light is like smoke, pale the way his dream was lit. He thinks about the dream, the brightness of the egrets in the grey rooms of the museum. Which way is it to London? It doesn't really matter, he has no intention of taking the direct route. London is 500 miles as the crow flies from Aberdeen, but before he has even crossed the border into England he will already have wandered this distance, following his curiosity wherever it leads him. There are things he needs to see along the way, plants and birds to catalogue, places he has never been. Also, he is reluctant to leave the mountains too soon. He knows that once he descends out of them, on the long haul to London, the mountains will wrench at him terribly. So he pulls his long blue coat over his back and starts walking into the deep mountains to the west of Aberdeen. Already he feels his mind thawing; in his journal he writes, *I am at length free*. By the time he staggers into London, six weeks and 838 miles later, the blue of his coat will be weathered with grey like the plumage of the merlin that brushed past him in pursuit of the lark that day.

There are fifteen breeding diurnal birds of prey found in the British Isles. This list does not include boreal migrants – bearing news from the Arctic – like the rough-legged buzzard and gyrfalcon, or rare vagrants such as the red-footed falcon, who occasionally brush the shores of these islands. Neither does it include owls. For they are raptors too; that is, a bird possessing acute vision, capable of killing its prey with sharp, curved talons and tearing it with a hooked beak, from the Latin *rapere*, to seize or take by force. But owls belong to a separate group, the *Strigiformes*. And although the change of shift between the diurnal and nocturnal birds of prey is not always clear-cut (as I experienced with short-eared owls on Orkney), owls require a list of their own; they are such a fascinating, culturally rich species, they need to be attended to in their own right.

Acute vision is a distinguishing characteristic of raptors. Just how acute is illustrated by this vivid description of a golden eagle recorded by Seton Gordon in *The Golden Eagle: King of Birds*:

> Four days later I had an example of the marvellous eyesight of the golden eagle. The male bird was approaching at a height of at least 1,500 feet. Above a gradual hill slope where grew tussocky grass, whitened by the frosts and snow of winter, he suddenly checked his flight and fell headlong. A couple of minutes later he rose with a small object grasped in one foot. It was, I am almost sure, a field-mouse or vole. Since he had caught his prey at an elevation well above that of the eyrie, he was able to go into a glide when he took wing and made for home. When he had grasped his prey he had torn

> from the ground some of the long grass in which his small quarry had been hiding: during his subsequent glide, as he moved faster and faster, the grass streamed out rigidly behind him.

Avian eyes are huge in relation to body size, and this is especially the case in birds of prey; many raptors have eyes that are as large, often larger, than an adult human's. The foveal area in the retina of birds of prey is densely packed with photoreceptor cells. A human eye contains around 200,000 of these cells; the eye of a common buzzard, by comparison, has roughly one million of these rod-and-cone photoreceptors, enabling the buzzard to see the world in much greater detail than we can. Images are also magnified in a raptor's eye by around 30 per cent. The birds' eyes are designed much like a pair of binoculars: as light hits the fovea pit in the retina, its rays are bent – refracted – and magnified onto the retina so that the image is enhanced substantially. Birds of prey see the whole twitching world in infinite, immaculate detail.

All fifteen of the diurnal birds of prey breed in these islands, though some in very small numbers. Many are permanent residents. The osprey, hobby, Montagu's harrier and honey buzzard are summer visitors. All are classified within a single order, the *Accipitriformes*, and subdivided into three suborders. *Accipitridae*: the soarers and gliders, the nest builders, distinguished by their broad 'fingered' wings; so: hawks, buzzards, eagles, kites and harriers. *Pandionidae*: with its solitary member, the osprey – a specialist – the hoverer-above-water, the feet-first-diver after fish. *Falconidae*: the speed merchants (whose nests are scrapes or squats): kestrel, merlin, hobby, peregrine; fast, agile fliers with pointed wings, capable (though not all of them do) of catching their prey in the air.

Pandionidae
Osprey

Accipitridae
Honey Buzzard
Red Kite
Sea Eagle
Marsh Harrier
Hen Harrier
Montagu's Harrier
Goshawk
Sparrowhawk
Buzzard
Golden Eagle

Falconidae
Kestrel
Merlin
Hobby
Peregrine Falcon

Fifteen birds of prey, fifteen different landscapes. A journey in search of raptors, a journey through the birds and into their worlds. That is how I envisaged it. The aim simply to go in search of the birds, to look for each of them in a different place. To spend some time in the habitats of these birds of prey, hoping to encounter the birds, hoping to watch them. Beginning in the far north, in Orkney, and winding my way down to a river in Devon. A long journey south, clambering down this tall, spiny island, which is as vast and wondrous to me as any galaxy.

Rain over the Pentland Firth. The cliffs of Hoy streaked with rain. The red sandstone a faint glow inside the fret. The low cloud makes the cliffs seem huge, there is no end to them. We could be sailing past a great red planet swirling in a storm of its own making. I am on the early morning crossing from Stromness to Scrabster. Light spilling from slot machines; the bar opening up; breakfast in the ferry's empty café. Through the window: an arctic tern, so beautifully agile, it seemed to be threading its way through the rain's interstices. Then a couple from Holland come in, hesitate when they hear themselves in the café's emptiness. They have been walking for a week through Orkney and their faces are red with wind-burn. They hunch over their breakfasts and I see how we do this too: mantle our food, like a hawk, glower out from over it. We are passing The Old Man of Hoy and all three of us shift across to the port side for a better view. We see the great stack in pieces, its midriff showing through a tear in the cloud. Then the rain thickens like a shoal and The Old Man, the cliffs of Hoy, dissolve in rain.

The way I'd pictured it, back home, doodling over maps, Orkney would be all hen harriers. Then the ferry, the train from Thurso, a request stop and stepping off the deserted platform into the blanket bog of the Flow Country. Then the vast, impossible search for merlins. I knew the birds were out there somewhere, not in large numbers, but I had seen a merlin once before, a skimming-stone, hunting fast and low far out on The Flows, the peaks of Morven, Maiden Pap and Scaraben on the southern horizon, a patch of snow on Morven's north face like the white dab on a coot's forehead.

All of that happened as best it could. On the crossing over from Orkney I thought of home – ached for it – for my family there, and thought of the fishermen out of Wick and Scrabster who, should they dream of home, hauled in their nets and headed back to harbour, not willing to tamper with a dream like that. I shared a taxi with the Dutch couple from Scrabster into Thurso. They were tired and polite and wanted to pay, Orkney's wind still rushing in them. They sat in the back of the cab, their faces glowing like rust.

Waiting for the train at Thurso, a slow drizzle, the rails a curve of light, glinting like mica in the wet. The last stop, as far north as you can go. Buffers, then a wall and then another wall because, if you did not stop, the train would slip like a birthing ship down through Thurso's steep streets, past her shops and houses and out into the frantic tides of the firth, rousing wreckers from their sleep who go down to poke about the shore like foraging badgers.

Then the warmth of the train, people steaming in their wet clothes. The start of my long journey south. All the staging posts between the moors of Orkney and the moors of Devon lying in wait for me. Reading the maps of each place obsessively, thinking the maps into life, imagining their landscape, their weather. The more I read the maps the more I imagined the possibility of raptors there. That hanging wood marked like a tide line above the valley: perfect for red kites. That cliff on the mountain's south face: surely there must be peregrines there … The train now pulling away from the north coast. A last glimpse of Orkney shimmering behind us in her veil of rain. Her harriers grounded, hunched under the dripping sky, feathers beaded with rain. The mark I'd made in the heather beginning to fade. Time on the train to reproof my boots, because the place I'm heading for, the next stop on my journey south, hints in its name that I should really be donning waders, flippers, a bog snorkel …

'Flow', from the Norse *flói*, a marshy place. The Flow Country, or The Flows as it is usually known, is the name given to the area of West Caithness and East Sutherland covered by blanket bog (literally bog 'blanketed' by peat). It is one of the largest, most intact areas of peat bog in the world, extending to over 4,000 km². Flick the noun into a verb and you also have what the landscape wants to do. The Flows want to flow, to move. The land here is fluid, it quakes when you press yourself upon it. The Flows is the most sensitive, alert landscape I know. A human cannot move across it without marking – without hurting – the bog. The mire feels every footprint and stores your heavy spoor across its surface. But it is a wonder you can move across it at all. There are more solids found in milk than there are in the equivalent volume of peat. The bog is held in place only by a skin of vegetation (the acrotelm), predominantly sphagnum, which prevents the water-saturated lower layer of peat (the catotelm) from starting to flow. And, oh, how it wants to flow! Think of the bog as a great quivering mound of water held together like a jelly by its skin of vegetation and by the remarkably fibrous nature of the peat. Think of that great mound breathing like a sleeping whale. For that is what it does. The German word is *Mooratmung* (mire-breathing). It is the process by which the bog swells and contracts through wet and dry periods. The bog must breathe to stop itself from flowing away.

As it breathes the bog changes its appearance. Unlike a mineral soil where the shape of the land is determined by physical processes, the patterns and shapes of the bog are continually shifting; peat accumulates and erodes, the bog swells and recedes. Occasionally, after exceptionally heavy rain, the water in the bog swells to such a volume that the peat, despite its great strength, can no longer hold the mound together. So the bog bursts, hacking a great chunk of itself away. In Lancashire in the mid-sixteenth century the large raised bog of Chat Moss burst and spilt out over the surrounding countryside, taking lives and causing terrible damage, a great smear of black water blotting out the land. Huge chunks of peat which were carried down the river Glazebrook were later found washed up on the Isle of Man and as far as the Irish coast.

The train follows the river Thurso. Herons in their pterodactyl shadows. The river so black it could be a fracture in the earth's crust, an opening into the depths of the planet. Passing Norse farmsteads, *Houstry*, *Halkirk*, *Tormsdale*. The Norse language here flowing down from Orkney and spreading up the course of the river. Flagstone dykes marking field boundaries. Sheep, bright as stars against the pine-dark grass, disturbed by the train, cantering away like brushed snow. Then the train is crossing an unmarked border, a linguistic watershed, the last Norse outpost before the Gaelic hinterland of the bog, a ruined farmstead with a Norse name, *Tormsdale*, peered down on by low hills, each one attended to by its Gaelic, *Bad á Cheò*, *Beinn Chàiteag*, *Cnoc Bad na Caorach*. The last stop, as far as the Norse settlers would go. Because if you didn't stop here you would wander for days adrift on the bog before you sank, exhausted, into the marshy *flói*.

'Bog bursts', 'quaking ground', 'sink holes' ... I am trying to pay heed to the dramatic vernacular of the bog, checking my boots are well proofed, my gaiters are a good fit. I must remember to check and recheck compass bearings against the map, must remember to tread carefully over this landscape. Because once, I nearly lost my brother in a peat bog.

Another family holiday, another peaty, midge-infested destination. This time, the Ardnamurchan peninsula,

Scotland's gangplank, the jump off to America. My brother was six or thereabouts and we had been to the village shop, where he had bought a toy car. More than a car, it was a six-wheeled, off-road thing. Orange plastic like a street lamp's sodium glare, round white stickers for headlights, a purple siren on the roof. My brother took it everywhere. And lucky for all of us he had the car with him on a walk up the hill one afternoon, holding the toy out in front of him, chatting to it, running through some imagined commentary, when he dropped, as if down a hole, into what looked like nothing more than just a puddle. The bog had got him. And he was struggling, sinking into the mire. But his instinct was to protect the car, to stop it getting muddy. So he held it out in front of him and by holding his arms out like that he stopped himself sinking any further and we leant over and hauled him out, oozing with black peat, like some urchin fallen down a chimney.

The train glides across the flow. Fences beside the track to hold the drifting snow. Tundra accents: greylag, skua, greenshank, golden plover. Wild cat and otter's braided tracks. Sphagnum's crimson greens. Red deer, nomads in a great wet desert, stepping between the myriad lochans. Mark the deer, for they can blend into the backdrop of the flow as a hare in ermine folds itself against the snow.

Then the train is pulling into Forsinard, where the Norse language has flanked around the bog and found an opening to the south through the long, fertile reach of Strath Halladale. And halfway down the strath met and fused with Gaelic making something beautiful. *Forsinard*: 'the waterfall on the height'. *Fors* from the Norse (waterfall); *an* from the Gaelic (of the); *aird* from the Gaelic (height). Clothes still dank with Orkney rain, smelling of rain, I stepped down off the train, crossed the single-track road, and walked out into the bog in search of merlins.

What else does he have in his knapsack, his machine? He has taken it off while he pauses to rest at Banchory, 23 miles out of Aberdeen. He leaves the road, sheds his coat and washes his hands and feet in the river Dee. He notices how people's accent here has slipped away from Aberdeen, a softening in the tone, a slower pace to it, as if the dialect here still carries a memory of Gaelic. And sure enough, a little further up the road he passes two men on horseback talking in Gaelic. He speaks Gaelic himself, has considerable knowledge of Scots and its many dialects. All along his journey he passes through the ebb and flow of dialect. Every mile along the road accents are shaved a fraction. Often he struggles to make himself understood. He might follow a seam of Gaelic like a thin trail through the landscape until it peters out on the outskirts of a town.

What else is in his knapsack? Two black lead pencils; eight camel-hair pencils with stalks; an Indian rubber; a shirt; a false neck; two pairs of short stockings; a soap box; two razors; a sharpening stone; a lancet; a pair of scissors; some thread; needles. In a small pocket in the inner side of his flannel undervest there is nine pounds sterling in bank notes. One pound in silver is secured in a purse of chamois leather kept in a pocket of his trousers. In all, ten pounds to last him through to London.

That first day he walks as far as Aboyne, 30 miles from Aberdeen. That night, at the inn, he writes in his journal until the candle has burnt down. He writes a long list of all the plants he has seen that day, both those in and out of flower. He dreams again of the museum, the place obsesses his dreams. But this dreaming is inevitable because the museum is the reason he is making this walk to London. He has heard that the British Museum holds an astonishing collection of beasts and birds, of all the creatures that have been found upon the face of the earth. And he *must* go to London to see these things. There are gaps in his knowledge, in the survey of himself, he needs to fill. As a student at Aberdeen he studied medicine for nearly five years, then, in 1817, switched to zoology. Since then he has devoted himself completely to studying the natural world. Linnaeus and Pennant have been his guides but now he has reached the point where he needs to set what he has learnt of the natural world against the museum's collection. He wants to check his own observations and theories against the museum's. Above all, he wants to see the museum's collection of British birds. Birds are what stir him more than anything. He is anxious to get there, to get on with his life.

* * *

The way I'd pictured it, back home planning this journey, was a neat transition: Orkney's hen harriers followed by merlins out on The Flows. Instead, on Orkney, merlins had darted through my days, led me astray across the moor in search of them. Then, not far out of Forsinard, in a large expanse of forestry, the first bird I saw was a male hen harrier, a shard of light, hunting the canopy.

Before I set out on this journey I had planned to try to look for each species of raptor in a different place, to dedicate a bird to a particular landscape, or rather the landscape to the bird, to immerse myself as much as I could in each bird's habitat. But the plan unravelled soon after I lay down in the heather on Orkney and a jack merlin, a plunging meteor, dropped from the sky, wings folded back behind him, diving straight at a kestrel who had drifted over the merlin's territory. It was astonishing to see the size difference between the birds, the merlin a speck, a frantic satellite, buzzing around the kestrel. He was furious, screaming at the kestrel, diving repeatedly at the larger bird until the kestrel relented and let the wind slice it away down the valley.

I stayed with that jack merlin for much of the day. Sometimes I would catch a flash of him circling the horizon or zipping low across the hillside, full tilt, breakneck speed. The sense of sprung energy in this tiny bird of prey was extraordinary, a fizzing atom, bombarding the sky.

Once I tracked the merlin down into a dusty peat hollow below clouds of heather. I marked the spot and started to walk slowly down the moor towards him.

Grandmother's footsteps: every few paces I froze and watched him through my binoculars. At each pause the colours of his slate-blue back grew sharper. Even at rest he was a quivering ball of energy, primed to spring up and fling himself out and up. Relentless, fearless, missile of the moor, you would not be able to shake him off once he had latched on to you. There are stories of merlins – like William MacGillivray's account of watching a merlin pursuing a lark amongst farm steadings and corn-stacks – where the falcon is so locked in on the pursuit of a wheatear, skylark, stonechat, finch or pipit (the merlin's most common prey species) it follows them into buildings, garages, in and out of people's homes. Even a ship out in the Atlantic, 500 miles west of Cape Wrath, became, for a week, a merlin's hunting ground. The crew reported that the merlin – on migration from Iceland – hitched a ride with them, chasing small migrant birds all over the ship, darting across the gunwale, around coiled hillocks of rope, perching on the bright orange fenders.

They don't always get away with it, this all-out pursuit; merlins have been known to kill themselves, colliding with walls, fences, trees. Merlins need the space – the sort of space there is on The Flows – to run their prey down. They do not possess the sparrowhawk's agility to hunt through the tight landscape of a wood.

I am still playing grandmother's footsteps with the merlin, but I do not get very far before one of the short-eared owls overtakes me and swoops low over the merlin, disturbing him, dusting him, so that he flicks away out of the peat hollow and lands again further down the slope. I mark his position. This time he has landed on a fence post. There is a burn running down towards the fence and I drop into it and use its depth to stalk closer to the merlin. I crawl quietly down the burn and, when I poke my head up again over the bank, the merlin is still there and I am very close to him. He is looking up at the sky, agitated. His breast is a russet-bracken colour, his back a blue-grey lead. That's it: he is gone. The female is above us, calling to him, a sharp, pierced whistle. And then I see the merlin pair together. She is a darker shape, a fraction larger. In his description of the merlin, William MacGillivray picks out the distinction between the male and female's dorsal colouring brilliantly. The male's upper parts he describes as a *deep greyish-blue*; the female's as a *dark bluish-grey*. But now I cannot make out any difference between the pair. Both the male and female merlin are gaining height, moving away from my hideout in the burn, pushing themselves into speed.

They were beautiful distractions, those Orkney merlins, pulling me after them, away from the hen harriers I was supposed to be watching. And throughout my journey, at every juncture, different species of raptor, inevitably, moved through the places I was in. So, hen harriers spilt out of Orkney and, like the Norse language, followed me across the Pentland Firth; merlins flickered through many of the moorland landscapes I visited; buzzards were present almost everywhere I went, however much I tried to convince myself they might be something else, the something that was eluding me –

goshawk, honey buzzard, golden eagle ... Every buzzard I saw made me look at it more carefully.

Quickly the map I had imagined for my journey became a muddled thing, transgressed by other birds of prey, criss-crossed by their wanderings. And though each staging post was supposed to concentrate on a single species, I loved it when I was visited, unexpectedly, by other birds of prey. I liked the sense that the different stops along my journey started to feel linked up by the birds, I liked the ways they set my journey echoing. Sometimes I came across a bird of prey again far from where I had first encountered it: a merlin on its winter wanderings in the south of England, an osprey on the cusp of autumn refuelling on an estuary that cut into Scotland's narrow girth.

The great Orkney ornithologist Eddie Balfour discussed, in one of his many papers on hen harriers, the minimum distance harriers nest from each other (hen harriers, notably on Orkney, will often nest in loose communities). But in a lovely afterthought to this, like a harrier pirouetting and changing tack, he touches on the optimum distance between nests as well, the distance beyond which breeding stimuli would diminish, neighbourly contact become lost. Extending the thought outwards from hen harriers living in a moorland community, he imagines larger raptors, golden eagles, with their vast, isolated territories, living, in fact, like the harriers, within a single community that extended across the whole of the Highlands, each nest within reach of its neighbour, like a great network of signal beacons.

I was fascinated by this idea of a community of raptors extending right across the country. It touched on my experience of re-encountering and being revisited by birds of prey as I journeyed south. Balfour's idea also seemed to challenge the notion that many birds of prey were solitary, non-communal predators, inviting the idea that even a species we perceive as being fiercely independent, like the eagle, still belongs to – perhaps needs – a wider community of eagles. It got hold of me, this idea, it got hold of the initial map I had sketched for my journey and redesigned it. Instead of moving from one isolated area of study to the next, from Orkney to the Flow Country and so on, I started to see myself passing through neighbourhoods – through communities – of raptors, the boundaries of my map – the national, topographic, linguistic borders – giving way to the birds' network of interconnecting, overlapping territories. A journey *through* birds.

The first bird I see as I am walking through the forestry above Forsinard: a male hen harrier, hunting the sea of conifers. And I could still be on that hillside in Orkney, except, what has changed? The bird, the bedrock, remain the same, but the sky is different here, not always rushing away from you as it is on Orkney. The wind is not as skittish here, the vastness of the land seems to stabilise it, give it traction. On Orkney I wonder if the wind even notices the land. And what else has changed, of course, are the trees that no more belong out here on the bog than they do on Orkney, where trees don't stand a chance against that feral wind. But still there are conifers here planted in their millions, squeezing the breath out of the bog. And today the male harrier is hunting over the tops of the trees just as I watched the Orkney harriers quartering the open moor. It is the same procedure except here, over the forestry, he is looking for passerine birds to scoop out of the trees. It is fascinating to watch the harrier hunt like this, as if the canopy were simply the ground vegetation raised up by 20 feet.

To begin with the newly planted forests would have been harrier havens, just the sort of scrubby, ungrazed zones they like to hunt, ripe with voles. But all of that is gone once the trees thicken and the canopy closes over, suffocating the bog. Greenshank, dunlin, golden plover, hen harrier, merlin, birds of the open bog, are forced to move on, or cling on, as this harrier was doing, trying to adapt to his changed world. Recently, hen harriers – always assumed to be strictly ground-nesting birds – have been observed nesting in conifer trees where the plantations have swamped their moorland breeding grounds. Hopeless, inexperienced nest builders, little wonder their nests are often dismantled by a febrile wind.

But this harrier I am watching over the forest still has its nest on the ground. From my perch on the hillside I draw a sketch of his movements over the trees. Meandering, methodical, he covers every inch of the canopy. I watch him drift above the trees like this for half an hour until (I recognise it from Orkney) there is a sudden shift in purpose to his flight. He stalls low above a forest ride, hesitates a fraction, then whacks the ground with his feet. I make a note of the time: 14.50: he lifts from the kill and beats a heavy flight direct across the tops of the trees; there is the female harrier rising towards him; 14.51: the food pass; 14.52: the female keeps on rising, loops around the male; 14.53: she goes down into a newly planted corner of the forest. The trees are only a few feet tall here and I mark the position of her nest: four fence posts to the right of the corner post, then 12 feet down from the fence.

The hen harrier is the bird that brought me to William MacGillivray. The moment came when I was meandering, harrier-like, through books and papers, field notes and anecdotes about hen harriers. Then I read this passage from MacGillivray's 1836 book, *Descriptions of the Rapacious Birds of Great Britain*:

> Should we, on a fine summer's day, betake us to the outfields bordering an extensive moor, on the sides of the Pentland, Ochill, or the Peebles hills, we might chance to see the harrier, although hawks have been so much persecuted that one may sometimes travel a whole day without meeting so much as a kestrel. But we are now wandering through thickets of furze and broom, where the blue milkwort, the purple pinguicula, the yellow violet, the spotted orchis, and all the other plants that render the desert so delightful to the strolling botanist, peep forth in modest beauty from their beds of green moss. The

golden plover, stationed on the little knoll, on which he has just alighted, gives out his shrill note of anxiety, for he has come, not to welcome us to his retreats, but if possible to prevent us from approaching them, or at least to decoy us from his brood; the lapwing, on broad and dusky wing, hovers and plunges over head, chiding us with its querulous cry; the whinchat flits from bush to bush, warbles its little song from the top-spray, or sallies forth to seize a heedless fly whizzing joyously along in the bright sunshine. As we cross the sedgy bog, the snipe starts with loud scream from among our feet, while on the opposite bank the gor cock raises his scarlet-fringed head above the heath, and cackles his loud note of anger or alarm, as his mate crouches amid the brown herbage.

But see, a pair of searchers not less observant than ourselves have appeared over the slope of the bare hill. They wheel in narrow curves at the height of a few yards; round and round they fly, their eyes no doubt keenly bent on the ground beneath. One of them, the pale blue bird, is now stationary, hovering on almost motionless wing; down he shoots like a stone; he has clutched his prey, a young lapwing perhaps, and off he flies with it to a bit of smooth ground, where he will devour it in haste. Meanwhile his companion, who is larger, and of a brown colour, continues her search; she moves along with gentle flappings, sails for a short space, and judging the place over which she has arrived not unlikely to yield

> something that may satisfy her craving appetite, she flies slowly over it, now contracting her circles, now extending them, and now for a few moments hovering as if fixed in the air. At length, finding nothing, she shoots away and hies to another field; but she has not proceeded far when she spies a frog by the edge of a small pool, and, instantly descending, thrusts her sharp talons through its sides. It is soon devoured, and in the mean time the male comes up. Again they fly off together; and were you to watch their progress, you would see them traverse a large space of ground, wheeling, gliding, and flapping, in the same manner, until at length, having obtained a supply of savoury food for their young, they would fly off with it.

Attentive, accurate, warm and intimate, you cannot help but feel MacGillivray's delight at being out there amongst the birds. The degree of observation: the way he records the hen harrier's flight, the detail in the landscape, the description of the moorland flowers and moorland birds. It felt to me like the work of an exceptional field naturalist; the writer seemed to notice everything. And I wanted to read more, had to read more. Felt a kinship there, at least in the way MacGillivray responded to the birds. He caught the hen harrier's beauty in his careful, graceful writing.

* * *

I walk through the middle of the day across the bog. Anything that breaks the horizon draws you towards it. The house is so far out on the flow it is like a boat set adrift. Not long abandoned, the building sagging, tipping into the bog. A portion of the corrugated-iron roof torn back, exposing timber cross-beams. Rock doves blurt out of the attic. Outside the house there is a bathtub turned upside down; four stumpy iron legs sticking upright, like a dead pig. In one of the rooms: a metal bed frame, a mattress patterned with mildew, blue ceramic tiles decorating the fireplace. A dead hind in the doorway, the stench of it everywhere. Deer droppings piled against the walls as if someone had swept them there.

MacGillivray often slept in places like this on his long walk to London. One night, on the outskirts of Lancaster, tired and wet, he stumbled upon a large, misshapen house in the dark. He went inside and groped his way around till he had made a complete circuit of the rooms. There was no loft, not even a culm of straw to bed down in, but earlier he had tried to sleep under a hedge and the house, despite its damp clay floor, was preferable. So he slept behind the door in wet feet with a handkerchief tied around his head, woke to a mild midnight to peep at the moon and walk up and down the floor a bit. Then slept again with his head on his knapsack to wake at dawn and walk down to the river to wash his face.

But that restless night on the outskirts of Lancaster comes much later. It is only five days since MacGillivray set out from Aberdeen and he is still in Aberdeenshire, crossing the Cairngorms on route from Braemar to

Kingussie. He spends the night of Sunday 12 September high in the mountains in a palaver of sleeplessness and shivering. Supper is a quarter of a barley cake and a few crumbs of cheese. After which he does his best to make a shelter out of stones and grass and heath, then settles the knapsack and some heather over his feet, to try to keep the cold at bay.

He wakes at sunrise and resumes his climb into the mountains. It is slow going and he pauses often to rest; his muscles, after so little food, shiver with fatigue. At the source of the Dee he pauses to drink a glassful of its cold, blissful water. Up on the plateau he finds moss campion and dwarf willow; in the steep grey corries: dog's violet, smooth heath bedstraw, alpine lady's mantle.

During my time in Caithness and Sutherland I heard rumours of merlins. People were generous with their knowledge, pointing places out to me on my map. Somebody's faint memory of a nest site was enough to set me trekking far out across the bog. One afternoon I walked out to a distant mountain that rose sheer out of the flow. I had been told that a corrie high on the mountain's north face was a traditional nesting site for merlins. I walked there through a land creased with water, through dense clusters of *dubh lochans*, the hundreds of small lakes that form beautiful patterns across the surface of the bog. In some places the lochans are packed so tightly you can drift through their mazy streets for hours with no sense of where you might emerge. Most of the pools are shallow, two or three feet deep, though occasionally one would sink its depth into blackness. Water horsetail grew in some of the shallower pools, bell heather and cotton sedge along the banks. Around the edge of the pools were great mounds of sphagnum moss built up like ant hills. I pushed my arm into one of them, losing it up to my shoulder in the moss's cool dampness, sphagnum tentacles crawling over my skin. Some of these mounds had been perched on by birds, wisps of down feather left behind, the imprint of the bird's weight on the soft moss.

Hours I spent out there on the bog, and so many distractions on the way to the mountain, so much water to weave around. At one point, I gave up and slithered otter-like between the lochans, swirling up clouds of peat particles when I dived into the pools. And somewhere out on the flow a great boulder – just as the abandoned house had done – drew me towards it. A huge lump of rock, 20 feet high, jettisoned by the retreating ice. There was a solitary mountain ash growing up through a crack in the rock like a ship's mast. I clambered up the boulder and found, on the slab's flat top, a plate of bones. I had discovered an eagle's plucking perch, bones strewn everywhere, on the slab and in the heather around the base of the boulder. Amongst the bones there was a red deer's hoof, its ankle still clothed in grey hide.

It was mid-afternoon by the time I reached the mountain and climbed up to the corrie. I sat down with my back against a rock, listened and waited for the merlins.

A corrie is the mountain's cupped ear. It is a contained space away from the rushing noise of the tops, an amphitheatre of silence. You walk into it and enter an enclosed stillness where everything is suddenly closer, amplified, the raven's croaking echo, the golden plover's whistle. I was glad to be out of the wind. I thought, if merlins were here, their calls would sound cleaner, sharper, and hopefully I could track them more easily by listening out for the birds. But, something about the quiet stillness of the corrie, the release from the rushing wind out on the flow … when I sat down beside the rock, I fell asleep and when I woke the corrie had grown cooler, thicker with shadow.

Later I heard another rumour, a sure bet this time, a place where merlins nested year after year. The site was a deep cut through the flow where a burn wound down towards the river. I walked there across the glittering bog, light finding and lifting pools into pools of light. Near the burn I found signs of merlins everywhere: chalk-marked boulders – perches, lookout posts – patterned white by the birds' excretions.

There are some neighbourhoods of the moor that draw merlins to them time and again. It is difficult to identify what it is about a location that has so much appeal, but availability of prey and suitable nesting sites play a crucial role in the land's capacity to draw in raptors. In his pioneering study of merlins in the Yorkshire Dales published in 1921, William Rowan observed nineteen different pairs of merlins return to the same patch of heather every year for nineteen years. Each spring there was always a new influx of birds because each year, without fail, both the male and female merlin were killed by gamekeepers on their breeding ground and their eggs destroyed.

It was enough for keepers to set their traps on top of a merlin's favourite lookout boulder. No need to even camouflage the trap: the merlin's fidelity to certain perches always outweighed the bird's mistrust of the sharp jaws of a trap. Rowan used to plead – even tried to bribe – the keepers to spare the merlins. He had watched grouse quietly foraging bilberry leaves right in front of the adult merlins at their nest, so he knew that merlins posed no threat to grouse. But a few days before the grouse season opened the keeper would go out early with his gun and clear the moor of hawks of every shape and size. And every year another sacrificial pair of merlins arrived to plug the gap. Rowan wondered where they came from, this surplus tap of merlins, replenishing the same patch of moor year after year. What was it about that clump of heather on the side of the fell that had such a pull on the birds? Rowan identified a few characteristics of the place, of merlin nests in general: a bank of deep, old heather; an expansive view from the nest site of the surrounding moorland; a number of lookout boulders above the nest ...

But it is hard to see the land as the birds must see it, to feel a place as they must do. I am always looking for clues in the landscape, trying to anticipate the birds from the feel of a place. The ornithologist Ian Newton observed that, when he was studying sparrowhawks in South West

Scotland, he could glance inside a wood and tell straight away if its internal landscape was conducive for sparrowhawks. Eventually, after you have spent time amongst the birds, once you have settled into their landscapes, you can walk through a wood or sit above a moorland burn and think: this is a good place to be a hawk, I could be a hawk here.

I keep well back from the burn and settle myself against one of the boulders. The rock is limed with merlin droppings and I think, of all the things that draw a merlin down onto a particular bank of heather, these white-splashed boulders, like runway lights, must guide the birds in, signalling that this is a good place for them, signalling to the birds that generations of merlins have bred here.

Then – my notebook records the time – 10.30: 'Heard male merlin calling and turned to see him just above me. He gained height and then flew fast, dropping to ground level and skimming out across the bog, a smear of speed …' I try to keep up with him but he is rushing so low against the ground, eventually the haze and fold of the moor fold him away. I try to pick him up again, but the vast acreage of sky, the speed of the tiny bird … I have lost him.

That was the pattern for much of the day. There were long absences when the merlin was hunting far out on the flow. I caught the occasional flash of him through my binoculars when he glanced above the skyline, followed him as he tumbled downwards and levelled out over a great sweep of the bog. Sometimes I was impatient to follow him out across the flow, to try to intercept him out there. But I knew that would be hopeless, I could never follow a bird so absorbed in its own speed. Gordon would have kept still and waited. Rowan would have kept still; I thought of William Rowan, his night-long vigils in the cramped hide on Barden Moor, buried in tall bracken, so close to the merlins' nest that, when he lit a cigarette to help ward off the flies that infested his hide, the smoke drifted over the female merlin, parted round her, making slow eddies of itself as the falcon bent to feed her young.

You have so little time to take in what you see of merlins. Their world is glimpsed in snatches of blurred speed. I was lucky on Orkney to have spent time beside a merlin who paused long enough for me to notice the russet plumage of his breast. But the merlin seen rushing past you in a bolt of speed is just as beautiful. At times, from my perch above the burn in the Flow Country, watching the male merlin coming and going, it felt like I was in a meteor storm. Always I heard him calling first, then scrambled to pick him up just in time to glimpse his sharp-angled wings and his low rush up the burn. On one approach, I heard the male call and the female answered him, a high-pitched *cheo, cheo, cheo*. As she called (still out of sight) I noticed the male suddenly jink mid-flight as if he'd tripped over a rise in the moor. Then it looked as if he had grazed, scraped something – another bird – because there was a small explosion of feathers beneath him. And just at that moment I saw a second bird rising from under him and realised it was the female (who up till then I had not seen) meeting the male there, receiving prey from him.

Later, when the moor was quiet, I walked up to the spot where the merlin pair had met. When the female gathered the prey from the male she must have scuffed it, loosening feathers from the dead bird. There was a dusting of feathers across the site where the food pass had taken place: down feathers snagged in the heather and in the cotton grass. How else could two birds of such charged intensity meet except in an explosion, a fit of sparks? I picked up some of the loose feathers and lined my pockets with them. Then I walked on up the burn, following in the merlin's slipstream.

III

Golden Eagle

Outer Hebrides

Before the long walk to London, before university in Aberdeen, before birds had entered his life, William MacGillivray learnt to shoot a gun. The first shot he fired – his uncle supervising, leaning over him, telling him to keep the butt flush against his shoulder, to anticipate the recoil – he blasts a table, hurries up to it afterwards through the bruised air to inspect the splintered wood.

Getting his eye in, his uncle nodding, encouraging. With the second shot, he brushes a rock pigeon off the cliffs, flicking the bird into the sea. His third shot hits two pigeons simultaneously, sets them rolling like skittles. Recharging his gun with buckshot, peering over the cliff, searching for the pigeons bobbing in the thick swell. *Let the barrel cool, William.* His uncle saying this and as he says it MacGillivray tests the barrel with his finger and flinches at the heat, and in that instant the gun worries him, becomes something more physical, and his shoulder wakes to the ache of the recoil. So by the time he fires the fourth and fifth shots he is too wary of it, of the power of the thing, his body squinting, flinching when he squeezes the trigger. And he pulls the shots wildly. His uncle lightly ribs him at this: *You even missed the mountain!* Some hours later, when the barrel has cooled, when his anxiety has cooled, MacGillivray fires the gun again. This time he hits and kills a golden eagle.

The first time I fired a gun? I must have been the same age as MacGillivray was, ten or eleven, staying at a friend's house on a farm. We drove out with his father one evening in a pickup truck, bumping slowly along the side of a wood, scanning the brambly undergrowth. At first I could not find the rabbit, though my friend kept pointing to it, only a few feet from the truck; puffy, weeping eyes, hunched in on itself. A crumpled, shuffling thing. It had not noticed us. Its eyes looked like they had been smeared with glue. A mixy, his dad said; here, put it out of its misery. He turned the engine off, draped an old coat over the open window, a cushion for me to lean the rifle on. Then, like an afterthought, as if he felt it would stop the gun shaking in my hand, he pulled the handbrake up; its wrenching sound like a shriek inside the truck.

So MacGillivray waits for the ache in his shoulder to subside. Then sets about gathering what he will need to shoot the eagle: a white hen from his uncle's farmyard, some twine, a wooden peg, a pocketful of barley grain, newspapers, the gun. He walks out of the farm and up the hill. When he reaches the pit that has been dug into the side of the moor, he ties the hen's leg to the twine and fastens the twine to the wooden peg. He pushes the peg into the ground, sprinkles some of the grain beside the hen and primes the gun with a double charge of buckshot. Then he retreats to the pit with its roof of turf. When he enters the hide it is as if the moor folds him into itself. He can keep an eye on the hen from a peephole cut into the wall of the pit. He starts to read the newspapers. Rain seeps through the roof and the damp paper comes apart when he turns it. He finds himself rolling scraps of newspaper into balls of mush in his palm until it looks as if he is holding a clutch of tiny wren's eggs. Outside he can hear the hen shaking the rain off itself.

He is dozing when he hears the hen scream. He scrambles to the peephole and there is the eagle fastened to the back of the chicken. The eagle is so huge it has shut out the horizon with its wings. The hen looks as if it has been flattened. MacGillivray hurries to pick up the gun, takes aim, and fires. But he has overloaded it with shot and the recoil shunts him backwards, the butt smashing into his cheek. He cannot see what has happened – if he has hit anything – as smoke from the gun has engulfed the eagle and the chicken, so he pushes through the heather doorway of the hide and rushes up to the target. The shot, he sees, has entered the side of the eagle and killed the great bird outright. The hen, amazingly, is still alive, trying to hobble away. So this is an eagle, he thinks; it is *nothing wonderful after all* ... Gun in one hand, hen in the other, he throws the eagle over his back and brings its legs down on each side of his neck. Then he sets off back down the hill, wearing the huge bird like a knapsack.

Eagles started to make themselves felt while I was searching for merlins in The Flows of Caithness and Sutherland. There was the eagle's plucking post – that huge flat-topped boulder, littered with bones – I came across far out on the bog. Now and then, too, I saw eagles – often a pair – rising high over the mountains to the west. It was hard not to act on these sightings, to stay on the flow and not follow the eagles into the west. Once, watching the male merlin rising above the burn, in the far distance, perhaps a mile away, I saw an eagle circling high over the moor. I liked the symmetry in that moment, lining up the smallest raptor in these islands with the largest. A telescopic projection, the merlin's wingspan magnified onto the rising eagle, pointing the way to the next stage of my journey. So I left the merlins above their burn in the Flow Country and followed that eagle into the west, to spend a week amongst the mountains of Lewis and Harris, William MacGillivray's stamping ground, the place where he shot the golden eagle on that morning of steady drizzle.

What became of the white hen and the eagle that he shot? When MacGillivray came down off the hillside with the eagle slung over his shoulders the whole village came out to greet him. It looked, at first, as if he was carrying a bundle of heather on his back. *Surely the boy could not have shot an eagle?* MacGillivray's uncle proud as punch, one eagle the fewer to worry his lambs. And because then MacGillivray knew no better, because birds had still to enter his life, he did not count the eagle's quills, did not measure its bill, did not dissect it to see what it had eaten. Instead, the eagle was dumped on the village midden. The hen, miraculous escapee, lived on, reared a brood, was then eaten.

The villagers call MacGillivray *Uilleam beag* (Little William). He comes to live among them, on his uncle's farm on the Isle of Harris, when he is three years old. Before Harris, before he acquires his Gaelic name, there is not much to go on. MacGillivray's father, a student at King's College in Aberdeen, leaves soon after William's birth in January 1796, joins the army and is killed fighting in the Peninsular Wars. MacGillivray's parents are not married and his mother, Anne Wishart, is never mentioned again. So his birth is a hushed, awkward thing and the boy is bundled away to be brought up by his uncle's family on their farm at Northton in the far south-west corner of Harris. When he is eleven, MacGillivray leaves the farm for a year's schooling in Aberdeen before enrolling at the city's university. Always returning from Aberdeen by foot to Harris for the holidays; a journey of over 200 miles, walking across Scotland's furrowed brow, at ease within his solitude, hurrying to catch the Stornoway packet out of Poolewe, back home across the Minch.

After a day of rain and fuming cloud I found the eagles hunting at first light. I saw their shadows first, moving fast over outcrops of gneiss, the grey rocks glancing in and out of shade. The rush of their shadows betrayed the eagles; only storm clouds move that fast over the land. They were a pair and hunting low over low ground at the north end of the glen. I climbed a tall boulder, an uprooted molar, and perched on its flat top to watch.

The sun that morning was out of all proportion, suddenly huge and close. And what that sun did to the eagles ... It found their hunting shadows and, as it rose, lengthened them like ink spills along the bottom of the glen. It found and lit the crest of gold down the back of the eagles' necks in a clink and flash of bronze. And it lasted only a few seconds. But you hardly ever see that in a golden eagle, you are never close enough to see the bird's golden hue, to the extent that you wonder how the bird got its name at all, when most of the time all you see is a great dark bird which MacGillivray knew simply as the Black Eagle. But there it was below me, the huge bloodshot sun finding and lighting up the delicate golds brushed into the eagle's nape.

Bird of silence and the clouds, what sort of hunter are you? We tend to exaggerate you out of all proportion. Stories of you driving adult deer over cliffs, lifting human babies when their mother's back is turned, fights to the death with foxes, wildcats, wolves ... In the city where I work there is a swinging pub sign with an image of an eagle, talons clasped around a human child, flapping heavily, bearing the infant away through the cobalt blue. You must forgive us our silliness, our intolerance. For when we meet you your sheer size is dizzying. Seton Gordon once mistook a golden eagle for a low-flying aeroplane. Even MacGillivray, who so often saw eagles in the Outer Hebrides, was stunned when once, at the edge of a precipice, a huge golden eagle drifted off a few yards in front of him while a great mass of cloud rolled over the cliff. MacGillivray was close enough to almost touch the eagle and so struck by the bird's presence, he shouted out from the top of the cliffs: *Beautiful!*

In reality a golden eagle could never lift something as heavy as a human baby, let alone a child. Even a mountain hare often needs to be dismembered, a lamb broken in two, before an eagle can carry a piece of it away. Gordon once wrote to a Norwegian lady after he heard about a radio broadcast she had made recounting her experience of being carried away by an eagle as a child. Gordon wrote to the woman asking if he could quote her experience in his book. She replied saying that she might consider his request on payment of a £25 fee.

Up on the bealach there is a movement along the ridge to the west: long-fingered wings, an eagle gliding just a few feet above the ridge. Then another eagle joins it: the male, noticeably smaller than the female.

Close up, when he was observing eagles from a hide – as Rowan did with his merlins – Gordon could tell the difference between the male and female golden eagle by the compact tightness of the male's plumage. But at a distance it is hard to distinguish between the sexes until the pair come together, and then the size difference is obvious, as it is with many other raptors, particularly the falcons, hawks and harriers, where reversed size dimorphism (when the female is larger than the male) is a characteristic of the species, in contrast to non-raptorial birds where the male is usually the larger. Some of the male golden eagles MacGillivray saw over these mountains appeared so small he thought for a while that another distinct species of eagle existed in the Outer Hebrides, yet to be identified, cut off from the world in these remote hills.

Reversed size dimorphism is most pronounced in those raptors that hunt fast, agile prey. So, for example, the greatest size difference between the sexes is found in the sparrowhawk, where the female is nearly twice the weight of the male. By contrast, there is far less size variation between the sexes in those raptors that are insectivorous or feed on slow-moving prey, and no difference at all in carrion-feeding vulture species. Female raptors must substantially increase their fat reserves in order to breed successfully. This increase in weight does not equip them well to pursue and catch their quick, elusive prey. However, male raptors (those that hunt avian prey especially) must remain as small and light as possible in order to hunt successfully to provide food throughout the breeding season for the female and their brood. Size dimorphism is reversed in many birds of prey because the male cannot afford to become too big; he needs to maintain an efficient size and weight in order to hunt efficiently. The female can afford to become bigger because, during the critical breeding season, she does not hunt as prolifically as the male. This hunting respite allows her to lay down sufficient fat reserves to produce and then incubate her eggs. There is the additional advantage that the female's larger size better equips her to defend the nest against predators.

What a strange grey beauty these mountains have (MacGillivray compared them to *a poor man's skin appearing through his rags*). The land scraped bare, the moor a craquelure of gneiss. The warm rocks smoking in the rain. Like the earth must have been when it was raw and molten-new.

This morning I am watching a pair of golden eagles gliding low over the mountain into a strong headwind. Then they turn, so the wind is behind them, and drift away across the tops. Not once do I see them flap their wings. Wind-dwellers, they are at home inside the wind, can fly into a headwind as easily as they can fly out of it. There are accounts of eagles holding themselves motionless in wind so ferocious that men could not stand upright and slabs of turf were ripped from the rock and flung hundreds of yards.

I climbed after them. I wanted to try and follow this pair, to be up at the same height as the eagles, to be amongst them in their world, meet them as they skirted low across the ridges. For hours I walked along the tops of the mountains, red deer coughing their alarm barks at me. I sheltered from the wind in shallow caves amongst the boulder fields, sipping from my thermos, waiting for the eagles, daydreaming about finding MacGillivray's unknown species of eagle, his ghost eagle, undiscovered for centuries, the coelacanth of these Hebridean mountains. Spend long enough looking for eagles and you could be forgiven for being haunted by them. Sometimes I have gazed and gazed at a dark shape against the rock convinced it is an eagle, summoning that shape into a bird. Gordon once watched a pair of golden eagles fly into a passing cloud and followed their dim outlines through the depths of white vapour as though, he wrote, *they were phantom eagles, or the shadows of eagles*.

From one of my shelters I see the male golden eagle again, briefly, gliding back along the ridge towards me. Then he turns and shoots out over the deep glen with its black loch full of rushing clouds and now the reflection of an eagle rushing through those clouds.

* * *

Sometimes the hunt begins at a great height. Prey is picked out over a kilometre away – a rabbit cropping the machair, black grouse sparring at the morning lek. Wings are drawn back and the eagle leans into a low-angled glide. At the last moment, wings open, tail fans and talons thrust forward. Stand under the path of an eagle stooping like this and the sound of the wind rushing through its wings is like a sheet being ripped in two.

Golden eagles can also take birds such as grouse on the wing, pursuing the grouse in a tail chase. But eagles need a head start, they need to build up a substantial head of speed, coming out of a stoop for instance, to be successful in overtaking and seizing such a fast bird as the red grouse. Eagles often hunt in pairs, contouring low over the ground together and sometimes even pursuing prey together on the wing. They also, occasionally, hunt on foot, poking about for frogs, young rabbits hiding in the undergrowth.

More often, an eagle hunts only a few feet above the ground, patrolling the same airspace as the hen harrier and short-eared owl. Several times I have watched golden eagles (and once a pair of eagles) hunting like this, glancing over the land, hugging the contours, looking to trip over prey unawares. Stealth is a critical factor in the golden eagle's ability to catch prey. The bird's traditional upland prey across its northern European range, grouse and hares, are equipped with supreme agility and speed, and more often than not this speed enables them to escape a golden eagle. It's a wonder such a large bird could catch anything by surprise ... But when you watch an eagle hunting low across the hills it is hard to keep track of the bird. It blends its dark brown plumage against the hillside and clings to the peaks and troughs of the moor. Glide – pause – drop – strike.

I hang above the deep glen watching the surface of the loch. From my perch on the mountain I can see the wind moving across the water. Blocks of colour, like leviathans grazing inside the loch, shunt into each other, black into grey into blue. The surface ripples as if rain were fretting it. Then, two new shadows pass across the water: both eagles are now hanging over the glen and – I hear it first – a raven is beating out from the cliff to harry them.

I am used to the size of ravens. Besides buzzards they are by far the largest bird I see around my home. I often hear the raven's loud guttural croak as I hang the washing out, and I was glad to hear that call again up here in the mountains. But I had not expected the raven to be so shrunk beside the eagles. This great black bulk of bird I was so used to seeing and hearing, dominating the skies over my home, was utterly dwarfed, reduced to a speck of a bird, buzzing irritably around the eagle pair. I noticed how hard the raven had to work to keep up with the eagles. Either eagle could step away from the raven simply by folding its wings slightly and easing out of earshot of the raven who flapped and croaked, scolding after them. When the eagles pulled their wings back behind them like this, they grew falcon-like in their shape, tidying those great wings away, a split-second change of gear from soar to gliding speed. One of the eagles mock-swooped at the raven, suddenly rushing at it, effortlessly catching and overtaking the raven and then, at the last moment, glancing away.

No bird is so modest in its speed. There is not the impression of speed with golden eagles as there is with the more obvious sky-sprinters, the hobbies and merlins, who appear to live their lives through speed, though Gordon thought that the golden eagle was the fastest bird that flies, utilising its weight in a long glide to gain tremendous acceleration. Gordon once watched a male eagle descend to the eyrie from the high tops carrying a ptarmigan in one foot. The eagle was travelling so fast in its descent that it overshot the nest, rushing past it, before it swung round and was carried back up to the nest by its impetus. Gordon wrote that the speed of that eagle's dive was quite breathtaking and he calculated that the bird must have been travelling at around 2 miles a minute. An aeroplane pilot flying down the east coast of Greece recorded being overtaken by a golden eagle while the aircraft was travelling at 70 knots. As the eagle passed the plane at a distance of 80 feet the bird turned its head to glance at the aircraft before it eased past it at a speed, the pilot estimated, of 90 mph.

But usually it is difficult to gauge an eagle's speed and it is easy to think, because of its great size, that it is a heavy, laboured flier. Not until you see it glide over a wide glen in a single gulp or overtake a covey of ptarmigan flinging themselves down through a smoking corrie do you have a sense of what speeds a golden eagle is capable of achieving.

I watched the eagle pair and the raven skirmishing above the loch for fifteen minutes. The raven threaded between the eagles, barking at them, but it seemed half-hearted in its efforts to drive the larger birds off its patch, and there was almost a harmony in this dance, in the interaction between eagle and raven. The eagles seemed unconcerned by the raven jostling amongst them. Sometimes the eagles stole away from the raven and entered their own whirling dance, gliding towards each other and then a last-minute rush of speed as they sheared past, wingtips almost touching. Each time the eagles rushed at each other like this I was sure they were bound to collide, before the last-minute wing adjustment and the pair glanced past each other through the tightening air.

MacGillivray is awake, unable to sleep, running through the inventory of himself. Tomorrow he is planning another visit to the mountains on the border of Lewis and Harris, where he will be amongst friends, where his days will be overseen by eagles, where his unknown species of eagle shimmers on the edge of things. He sits up all night preparing for the trip. He was only in the mountains a few weeks ago but he wants to return to write – to finish writing – a poem. It has been nagging at him all winter, this poem. And he needs to go back to the mountains to get the details of the scenery right. The sense that he has too little material from his first trip to write the thing has been mithering him for months. But more than the poem this trip to the mountains is really about him trying to settle himself, to find a sense of resolution. He writes in his journal:

> The chief cause of all my disquietude is the want of resolution …

He needs to calm himself with purpose:

> At present I cannot help looking upon the vicissitudes of life with a kind of terror … If I had resolution, I should not despair …

And when he despairs his anxiety boils over and he turns his frustration mercilessly on himself:

> … such is the fickleness of my mind, that my whole life, hitherto has been nothing else than a confused mass of error and repentance, amendment and relapse … I am truly ashamed of myself, not to say anything worse …

But before he can begin his walk there is this long unquiet night ahead and all he can do is check the inventory of himself once more, rehearse his daily rituals, as if he were checking the contents of his knapsack:

- Rise with or before the sun.
- Walk at least five miles.
- Give at least half a dozen puts to a heavy stone.
- Make six leaps.

- Drink milk twice a day.
- Wash face, ears, teeth and feet.
- Preserve seven specimens of natural history (whether in propria persona or by drawing).
- Read the chapter on Anatomy in the Encyclopaedia Britannica.
- Read the Book of Job.
- Abstain from cursing and swearing.
- Above all procrastination is to be shunned.

Then it is dawn and MacGillivray is leaving for the mountains. He wants to reach Luachair, the tiny settlement at the head of Loch Rèasort where his friends live, before nightfall. But it is not the best of starts: he misjudges the tides and the great sands at Luskentyre are covered when he reaches them. Shortly afterwards, he loses his way in the mist and rain halfway up the ben.

North, then west, then north again, tacking up through the island. Tarbert to Aird Asaig, through the glen above Bun Abhainn Eadarra, up to the head of Loch Langavat. Lately the island does not recognise itself. People tipped out of their homes, the hills planted with sheep. And the villages all along the west coast of Harris, the machair lands at Nisabost, Scarista, Seilebost and Borve, all of them waiting to follow suit. Dismantling their homes, taking the roof beams with them to the new lives set aside for them on the island's east coast, a land so pitted with rock *not even beasts could live there*. Blackface sheep leaking out into the abandoned hills and glens, and so many of them now that, in the moment it takes to turn your back, it sometimes looks as if snow has crept back after the hills have thawed.

One night, descending the hill of Roneval, MacGillivray is spooked by a strange fire flickering on the hillside. He creeps to within a hundred yards of it but then holds back, skirting around the flames through the dark, convinced that the fire had been kindled by sheep stealers, rustling the sheep partly out of anger, partly to alleviate their hunger. *Dear Sirs, we are squatting under revolting conditions in hovels situated on other men's crofts* ... One hovel = two rooms = twenty-five people crammed to the rotten rafters in there. The people's situation beggars all description, the poverty just unimaginable; the shores scraped bare of shellfish, nettles, brambles ... uprooted and eaten, scurvy seen the first time in a century.

At the head of Loch Langavat, struggling through snow and snowmelt, MacGillivray turns north-west for Loch Rèasort. He has not eaten anything since he set out in the morning dark from Northton. Each bog, each bank of snow, seems to swell before him. But once he clears the watershed between Rapaire and Stulabhal, it is a long downhill glide to Luachair. There! He can see the thin glimmer of Loch Rèasort, the mountains are shutting down its light. Two hours after dark he reaches the house of his friends, opens the door, calls out a greeting to them.

* * *

Now I am looking down into the glen with its loch shaped like a rat. If I could bend down from this height and pick the loch up by its long river-tail it would squirm and wriggle beneath my grip. I'd heard of golden eagles doing that to adders, lifting the snakes and carrying them off in their talons as the snakes writhed and contorted in the air like a thread unravelling there. And more exotic things than snakes are sometimes taken by eagles. The lists of prey recorded are eclectic: grasshopper (Finland), pike (Scotland), tortoise (Persia), red-shafted flicker (USA), dog (Scotland, Estonia, Norway, Japan, USA), goshawk (Canada), porcupine (USA) ... More hazardous, perhaps, than even a porcupine was the stoat that an eagle near Cape Wrath was once seen to lift, the eagle rising higher and higher in a strange manner then suddenly falling to the ground as if it had been shot. The stoat had managed to twist its way up to reach the eagle's neck, where it fastened its teeth and killed the bird. Or the wildcat that an eagle was seen to lift in West Inverness-shire: the cat was dropped from several hundred feet and the eagle later found partly disembowelled with severe injuries to its leg.

The golden eagle is capable of predating a wide spectrum of birds, mammals and reptiles, and yet where possible they are essentially a specialist predator, feeding on a narrow range of prey items common to the eagle's mountain and moorland habitat. Golden eagles adapt to become a more generalist predator when their usual prey is scarce. In the Eastern Highlands of Scotland 90 per cent of golden eagle prey is made up of lagomorphs (mostly mountain hare, but also rabbits) and grouse (both ptarmigan and red grouse). In the Outer Hebrides their diet is more varied because their usual moorland prey, hares and grouse, are scarce here. So, in the Western Isles, rabbits, fulmars and, in winter, sheep carrion make up a high percentage of the eagle's diet. Grouse and hares are also less common in the Northern and Western Highlands than they are in the east, and consequently the eagle's diet in these regions is also varied, with deer carrion becoming important in winter. But as a general rule, carrion, despite its availability, tends to be much less significant in the eagle's diet through the bird's nesting season, when live prey are preferred, and tend to be easier to transport to the nest than bulky carrion items.

I settle down above the glen with my back against a boulder and keep watch. I can spend hours like this, waiting in the margins for the chance of birds. But today it is a long wait and I can feel the wind drying out my lips. I am just about to give up and move on when I see two golden eagles flying low down a steep flank of the mountain. Their great wings pulled back behind them, their carpal joints jutting forward almost level with the birds' heads. They cling so close to the side of the mountain they could be abseiling down the incline. An adult bird and a juvenile, the immature eagle with conspicuous white patches on the underside of its wings. All the time the young bird is calling to its parent, a low, excited *cheek cheek cheek*, the sound carrying down into the glen, skimming across the loch.

Both birds are only 100 feet now from where I am sitting. Through my binoculars I can see the lighter-coloured feathers down the adult's nape and the yellow in its talons. The juvenile's calling is growing more persistent, rising in pitch. Then the adult eagle drops fast into the heather with its talons stretched in front of it. At once the youngster is down beside its parent. Something has been killed there. The adult eagle rises and starts to climb above the glen. The young bird proceeds to dip and raise its head into the heather, wrenching at something. I stay with it, watching the young eagle feed. When it has finished it remains in the heather, carefully preening its left wing. Each time the young eagle lifts its wing there is a flash of white in the plumage like an intermittent signal from the grey backdrop of the mountain.

I realise I have witnessed something special, a juvenile golden eagle learning to hunt. The immature bird was clearly still reliant on its parents for food, but it was now piggybacking, tagging along while its parent hunted down the mountainside. The young bird flew so close to the adult, it must have seen the prey in the heather – whatever it was – at the same time as the adult eagle.

I remain watching the juvenile for the next twenty minutes and twice it lifts off the side of the mountain and lands again. Each time it lands it thrusts its talons hard out before it at the last minute as if practising its strike. The adult bird comes back into view again and the young eagle shoots up to meet it, circling its parent, crying repeatedly with its begging call, *ttch-yup-yup, tee-yup.* I have a very close view of the adult eagle as it swoops low across the cliff in front of me. Its tail is a dark grey colour, like the gneiss beneath it.

It all comes to a head, this gutting of the island from the inside out, when MacGillivray and his uncle are summoned by the laird to decide the fate of his uncle's farm at Northton. The laird's factor, Donald Stewart, is there, whom MacGillivray does not care for, whom he calls a wretch of a man. Donald Stewart, who is as ruthless as the sea, who will clear the bulk of the people from the west coast of Harris by 1830, who will even plough the graveyard at Seilebost till skulls and thigh bones roll about the ground like stones. And now Stewart has his eye on the farm at Northton, which is the finest farm in the country. So MacGillivray's uncle has been duly summoned to the big house at Rodel for this meeting – the Set, as it's known – and MacGillivray goes along to support him, to steady him, just as his uncle steadied MacGillivray when he was learning to shoot a gun all those years ago. The pair of them enter the house at Rodel where Stewart and the laird, MacLeod, are holding court at a large table, and before MacGillivray and his uncle have even sat down they are told that his uncle has lost the farm, that it has already been decided and would he like to bid for a different farm instead? What happens next is truly wonderful: MacGillivray's uncle, distraught and silent, listens, Stewart seethes and listens and MacLeod squirms and listens and gulps at his snuff, while MacGillivray stands up and harangues them, boiling over with indignation at the injustice, at Stewart's duplicity, at MacLeod's promises to him about the farm. MacGillivray is like the sea that night in January when the wind picked up the water in whirlwinds of agitation. He is furious with MacLeod and Stewart, but most of all he is furious with Stewart, who he knows is really behind all this, the wretch. And when he has finished fuming at them, MacGillivray sits down and there is a long silence. Stewart sits in cowardly silence and MacLeod, who is also a coward, gulps prodigiously at his snuff. Then MacLeod nods and it's settled and the farm is his uncle's for another year at least. The rent is set at £170, the meeting is over and MacGillivray and his uncle are both leaving and making straight for the nearest public house.

Seton Gordon once helped a golden eagle cross back over from death. He found the eagle hanging off a cliff in upper Glen Feshie. The bird's foot had been almost severed by the jaws of a trap which was fixed to the top of the cliff. Gordon and a companion quickly haul the eagle up the rock. They are shocked by how light the bird is, how many days it must have been hanging off the cliff. They carry the eagle, so weak it does not struggle, to some level ground. Very reluctantly, Gordon amputates the leg at the break. They wait and then watch the eagle as it starts to flop its way along the ground towards the precipice. Gordon's friend shouts at him to catch the bird before it is dashed to its death, but he is too late and the eagle has reached the edge of the cliff. It wavers there and they are sure it is going to fall to its death. But then the eagle tastes the icy uprushing current of wind, opens its broad wings, and rises.

The story of birds of prey in these islands is the story of the birds' relationship with humans. The nature of that relationship is integral to understanding the history and the current status of raptors in the British Isles. The status of golden eagles in the Outer Hebrides today is linked to the events that precipitated the meeting between MacGillivray, his uncle and the laird that day in April 1818. In particular, the eagles' current status is linked to the work of the factor, Donald Stewart, who wanted to evict MacGillivray's uncle and who went on after that meeting to clear the people from the west of Harris in the 1820s and 1830s to make way for his sheep. For the legacy of those clearances – the way those events have impacted the landscape of these islands – is still felt by the eagles today. The sheep that replaced the people brought, through poor husbandry, a ready supply of carrion to the hills, enabling golden eagles to live at high densities in these mountains. At the same time the intensive grazing pressure the hills were put under by the flocks (and more recently the large red deer population) produced, eventually, a degraded landscape, stripped of heather, of nutritional quality, and in turn denuded of its native herbivores, grouse and hares in particular. There have been dramatic declines across the region of grouse and hares, the golden eagle's natural prey. And whilst golden eagles can live at high densities in these mountains, they do not breed as well here. Live prey – its nutritional value – is critical to the breeding success of golden eagles; the birds breed better where live prey still thrives, as it does, for example, in the Eastern Highlands. The legacy of those clearances from MacGillivray's time is still felt, their impact on the land has been as dramatic as the ice.

I miss so much compared with them. Gordon, for instance, on the side of Braeriach, finding the perfect impression of a golden eagle's wings, like the imprint of a fossil, in the freshly fallen snow. MacGillivray: noticing how the currents in a channel are lit by different shades of brightness from the moon. Gordon: watching the aerial display of a male golden eagle, noticing that his neck seemed thinner than the female's and that the male's wings appeared larger in proportion to the body. MacGillivray: noticing the component parts of a raven's nest: heather, willow twigs, the roots of sea bent and lady's bedstraw ... MacGillivray: not noticing that the eagle he shot and slung over his back fitted him like a cowl, clung to him, its spirit clung on to him.

IV

Osprey

Moray Firth

Travelling north, early spring: the osprey's trajectory. March wakes a restlessness in the birds. Ospreys begin to part their winter quarters, leaving their roosts in the mangrove swamps of West Africa. Staggered departures, a slow release of birds. One morning rising higher than usual on the thermals above the palm-fringed coast, turning north instead of west in a mêlée of mobbing gulls.

What does that restlessness feel like in the bird, that longing to leave? It must be overwhelming. The muscles needing to work, tense for work. The pull of the north, the prospect of all that light, dragging at the bird like a rip tide. A thickening of need in the birds to be gone. The journey – the awful journey – thousands of miles across the Sahara, the Bay of Biscay, is not the point. It is getting back to spar and mate and claim a patch. Back to the deceiving north, still shuffling spring with winter.

The first ospreys make land off the south coast of England towards the end of March. Coming in over Poole Harbour, perhaps, up towards Rutland Water. The land below is lit by seams of water. Over the Lancashire Plain there is snow in the plough furrows and the crease of tractor prints, lighting the fields' veins like a sonogram. Crossing the Southern Uplands, past Tinto Hill, over the Clyde and the cold flanks above Arrochar. Reaching Loch Linnhe, the bird swings north-east and tracks along the watery motorway of the Great Glen. Between Loch Oich and Loch Ness the earth curves just enough for the osprey – the fish-hawk – to catch its first glimpse of the Moray Firth and the huge expanse of Culbin forest, the sea-black wood.

I am reading all I can about ospreys. Of all the dismal stories: the perceived threat to fishponds, the blanket intolerance ... The whole of England, Ireland and Wales rid of ospreys. Gone from England by the middle of the nineteenth century (though mostly by the end of the eighteenth). In the north of England the last known breeding attempts came at the end of the eighteenth century at Whinfield Park in Westmorland and the crags above Ullswater. A pair bred on Lundy in 1838 and on the north coast of Devon in 1842. The last known English nest site was at Monksilver in Somerset in 1847. By the beginning of the twentieth century there were only a handful of eyries dwindling in the Highlands. Lochs named after the osprey (*Loch an Iasgair*: the loch of the fisher) now outnumbered the birds themselves. And the osprey's rarity hurrying its demise: egg collectors hunting down the beautiful, valuable eggs. A last-minute rally to try to halt the inevitable: barbed wire is draped beneath the remaining eyrie trees, one nest is even given police protection. But the thieves stop at nothing: a man swims across a freezing loch at night, six inches of snow on the ground, clambering up the island's slippery castle ramparts to reach the nest, swimming back across the loch holding the eggs above the water, his accomplice hauling him in with a rope. Coming out of the black loch like that, his arms held above his head, he looks like a narwhal rising through a blowhole in the ice.

Two years earlier, May 1848, in north-west Sutherland, the egg collector Charles St John waits by the shore of Loch an Laig Aird while his companions go to fetch a boat. There is an osprey's nest on an islet in the loch and St John can just make out the white head of the female sitting on her eggs. He will write later that he is sorry for what he does next. Before the boat arrives, before they row out to rob the nest, the osprey rises and flies across the loch towards St John. When she comes into range he raises his shotgun and fires. Her wings are so long that when she tumbles to the ground it looks like a windmill has come unhinged and its blades are spinning, falling towards him. She falls slowly like an ash tree's rotating keys.

Late one May I headed north into the Highlands through a day of immense heat. Water-borne, water-navigating, following the course of the rivers, the Spey and then the Findhorn, travelling deep into osprey country. Rounding the Cairngorms, snow like a ferrule on the caps of the mountains.

I parked on the edge of Culbin forest and began to walk through the trees. Skin prickling with heat. Bronze-coloured pine seeds, like drops of resin, sticking to the sweat on my arms as if a vambrace was growing there. The forest quiet, sleeping off the day's heat. Wood ants turning over the dead pine needles. Cowberry plants beneath the trees, glossy leaves, bright amongst the granite-dark pines. Slender, silvery rowans like waymarks in the pine-gloom. The forest floor rippling, moving under the ant-pannage.

Ninety years ago Culbin was a restless desert, a vast area of shifting sands. A place so unlike anywhere else in Britain that at the end of the nineteenth century it became home to a population of Pallas's sand grouse, a species usually found in the deserts of central Asia. Culbin was known as 'Britain's Desert', large enough to lodge in the imagination, a place where history shuffled with mythology: skeletons excavated by the shifting sands were either travellers who had lost their way in its expanse or an ancient race who had lived and foraged along the shingle ridges.

I walked for an hour through the forest and came out of the trees onto the shore of Findhorn Bay. I swam and cleaned the heat from my skin. Mute swans glided past then rose and flew in low, heavy-wingbeat processions across the water. Herring gulls in their different shades of age. Oystercatchers in their black and white synchrony. The bay loud with birds, busy with wings, squabbles, tern-twisting flight. I waded back through the shallows and dried myself in the warm air. Late afternoon, I pitched my tent just inside the trees. From there I could lean out of the tent flaps and take in the whole sweep of the bay.

Autumn 1694, a violent snow-globe shake: a great storm blows in from the west. Sand pours down from the dunes and sweeps over the fields of Culbin like an avalanche. The sand rushes towards the men working in the fields. Ploughs are abandoned, sheaves of barley smothered. Everything is sealed in sand: orchards, cottages, the mansion house ... Whatever blocks the way is engulfed.

The sand has been working towards this for years, encroaching further and further east, deleting the land as it went. Sixteen seventy-six: the Culbin harvest is abandoned when the fields are drenched in sand. Year on year the estate keeps shrinking. The farms blotted out: Dalpottie, Laik, Sandifield, Middlebin. If you wanted someone to disappear to a remote spot the locals used to wish them, *Gang to Dalpottie!*

Culbin was destroyed in a single night, transformed into somewhere unrecognisable. The following morning the people woke in darkness, their windows and doors banked up with sand. There had been no warning, the storm coming out of nowhere, the sudden, rapid devastation. An act of God, they said: punishment meted out to an irresponsible landowner. After the storm the owner petitioned parliament to be exempt from cess (land tax), citing that the best three parts of his estate had been destroyed and what little remained was daily threatened by the dunes.

Then he is there, an osprey hovering over the bay. I am sitting at the edge of the forest, feet dangling over the crest of a dune. I found him a darker shape, hunting above the gulls, a high-hanging poise, a stillness there, so unlike the garrulous banter of the gulls. Long dark wings. Hovering, scanning, a kite-kestrel, 100 feet above the water. Then he drops, checks, hovers again. Then his drop tilts and gathers into a dive. In the last seconds, talons are thrown out in front of him to part the water and the osprey submerges in a great crashing plunge. A heartbeat later, he is rising, hauling the length of his wings out of the sea. Airborne, 10 feet up, he pauses, and what happens next is quite beautiful: he shakes the water from his feathers and a mist of droplets shivers off him like welding sparks.

I have never been so close to an osprey. I had watched one before, further south, in the Ochil Hills, in a buzzard-tangle over a reservoir. I heard the buzzard calling, looked up lazily thinking it was two buzzards sparring. But no, not a buzzard: look at the long, long wings, the carpal joints, angled sharp as Cuillin peaks. The buzzard was mobbing the larger bird, thumping down towards it, sending the osprey twirling and tumbling to avoid the buzzard's jostling bulk. For days after that encounter: wonder and exhilaration, to finally see an osprey, to grasp and store its shape.

He comes closer to the shore, drifting towards my seat on the dunes. I can see the black strip across his eyes. Head tilted downwards, reading the water. This time, when he dives, he is so close I hear the splash, a great cymbal crash of water and bird. I watch the talons go in and then come out again, clawing, empty: a miss. That grip! One fiftieth of a second is all it takes to clamp around the fish. The outer toe reversed so the fish is held in front and behind, clamped tight in the foot's sandpaper vice.

A grip like that would surely lead to speculation, stories, myths … And sure enough there are fishermen's tales of osprey skeletons found embedded in the back of fish too big to lift, fish who pulled the birds under, drowned them in the inky black.

There are stories too of young ospreys in the nest forced by their mother to gaze at the sun until the one that blinks first and weeps is dispatched, thrown out of the nest like some brutal Spartan ritual. And there are tales of fish turning their glistening bellies up, sacrificing themselves to the osprey as she cruises over their lake. For how else do you explain a bird so exemplary, so adept at plucking fish from out of the depths at will?

I watch him fall from a great height. Beneath him: a thimble-thin bay that fills and empties with grey, colloid water. At the last minute, before he enters the water, feet are flung forwards ahead of him. As he lines the fish up between his feet it looks as if his beak is reaching to touch the tip of his talons like somebody stretching to touch their toes. Then the splash! So close, this time, it almost wet me!

The turning point, the moment when ospreys returned to these shores to breed once more. When – how – did that occur? Some say they never completely disappeared, that a pair bred on in the quiet of the Caledonian forest. Gordon notes in his book *In Search of Northern Birds* that a pair of ospreys were seen beside a Highland loch in May 1938. One of the birds landed in a tree beside its mate with a fish, a good sign that they may have been nesting in the area, though Gordon is not so sure, suspecting the birds had simply paused a while on their migration to Scandinavia. And there were other sightings over the years, from Galloway to Ross-shire. An osprey was sometimes seen flying high over Loch Arkaig, one of the last places ospreys nested before they became extinct as a breeding species (1916 is regarded as the last year that ospreys bred successfully in Scotland, on Loch Loyne, Glen Garry). The few who knew of them kept their secret close, looked out for the birds each spring. Migrant birds were seen on passage to Scandinavia and small offerings must have been made to the gods by those who witnessed these migrations, willing the ospreys to rest a while longer, to stay. And in 1954 a pair did stay and raised two young from a Scots pine tree at the south end of Loch Garten. But word leaked out and in the following years the birds' eggs were frequently stolen. The turning point came in 1959 when the RSPB took the decision – a brilliant, bold decision – to advertise the ospreys' presence to the world. A public observation post was set up, telescopes trained on the nest. And visitors in their thousands streamed in to see the birds.

The burden of care for the ospreys became a public concern. The press reported the success or failure of breeding. Recolonisation was tentative, incrementally slow. By 1966 there were three osprey pairs in Scotland. Egg thieves still menaced, cutting through razor wire, climbing trees at night with the same determination as their forebear who had swum across the freezing loch with his stolen clutch. The thieves drastically hampered the progress of recolonisation, havocking the breeding season like a May gale. That awful journey, thousands of miles across the Sahara, the Bay of Biscay, for nothing. Once a nest had been robbed the ospreys would often build a 'frustration' nest somewhere nearby; too late in the season for rearing young, going through the motions, a grief nest. Volunteers monitored nest sites around the clock, number plates of suspicious cars were reported to the police. Some of the thieves were caught, some scared off. In 2000 a detachment of Marine Commandos dug themselves in for the night around an eyrie that had been repeatedly robbed. The following morning the Marines ambushed two thieves prospecting beneath the nest and marched them away to the police.

Egg theft is waning now, though not extinct. Custodial sentences, cultural shifts have seen to this. And ospreys are returning and returning; over 200 pairs breeding in Scotland and reaching, spreading out to England and to Wales, returning to their place-names in the landscape.

After the devastating storm of 1694, the sands that had covered Culbin continued to shift. Things that had been buried were suddenly revealed again. The chapel and the dovecot appeared like mirages adrift on the dunes before the sands shook and covered them once more. Trees emerged thinner, their trunks tapered where the sand had moved around them. On one occasion a chimney from the manor house peeked up out of a dune like a periscope. Whoever it was who came across the chimney there did what anyone else would have done, they clambered up the stoss, the windward slope of the dune, peered into the dark mouth of the flue and called out a greeting, *Hello down there* ... Whatever it was that called back at them from the bowels of the great house was surely more than just an echo. Otherwise, why did the man flee, terrified, careering down the side of the dune in a cascade of avalanches, the dune falling after him, repairing itself, smoothing over, as quickly as the man's tumbling descent gashed at the sand.

And once, or so the story goes, the tops of the laird's old orchard were revealed. And the trees blossomed so the sand became a quilt of pink and white. And in the autumn there were apples strewn across the dunes as if a game of billiards had been abandoned there.

I stayed in Culbin forest for two days and nights. In the forest I met treecreepers, tiny wren-sized birds, bibbed white, scratching up the pine trunks. I met a hare who lolloped past my tent one morning as I sat outside brewing tea. The hare turned to look at me but did not pause or alter its pace, just carried on threading its way through the forest understory. I met, as well, a roe deer, lustrous copper coat like a new two pence coin, grazing amongst the pine interstices, moving slowly through pools of shade. I sat very still and watched her until she almost grazed me. At the last minute scenting me, seeing me, springing, leaping away, stopping after 20 feet to turn and snort and stamp her foot.

On the north shore of the forest I met the most beautifully patterned rocks I have ever seen, as if the bedrock of the firth had been flensed and then spilt from inside it stones that had gestated for millennia. I wanted to pocket some of the rocks and take them home but they were so jewel-like and exquisite, so intricately textured, it felt like a theft to disturb them.

I heard among the trees a moaning, thinking it must be the wind in the tops. But then heard it again drifting off a sandbank and turned and saw sixty grey seals hauled up on a beach between the bay and the sea. Greys and whites and blues, skin patterned like barnacle-covered hulls, sighing, singing, wailing above the noise of the breakers.

And in between there was always the osprey. A noise like an intake of breath above my head and there he was coming in off the forest into the glare of the bay. Once I found him hovering high above the trees, so high up and so far away from the water I did not think he was hunting. But then a pause, a fracture in the hover, and he dropped into the narrow channel that links the bay with the firth. Splash! He rose with a fish grasped in his talons, the head of the fish facing the direction of flight, reducing wind resistance; a slight curvature, as if the fish was turning its head, where the talons gripped it.

Iasgair (the fisher); *Iolair-Iasgair* (the fisher eagle); *Iolaire an uisge* (the eagle of the water) ... The osprey has also been called the 'mullet hawk' as, in Scotland, grey mullet are so often taken by the birds. But in addition to mullet any fish that hunt or swim near the surface can be predated by ospreys: pike, carp, bream, trout, grilse, roach and perch ... In their tropical West African wintering grounds the variety of fish taken must be extensive.

Sometimes the osprey called as he flew, skimming over the treetops above me. *Chee-yup*, *chee-yup*: the first note, *chee*, long and slow. I looked up through the trees and there he was, directly above me, gliding over the rim of the forest, his white breast a drifting moon, a spotlight passing over me as I shrank down among the pine roots.

Often I found him between elements: on the cusp of the forest and the sea; coming out of a dive, ascending from the water back into the air. And I found myself drawn to these margins, looking for him there, poking about the shallows of the bay or sitting on the edge of the forest where the pines had lost their footing and fallen, like the man fleeing the ghostly echo, down the face of the dunes.

The lull of the afternoon, watching the tide glut the bay. No sign of the osprey. A distant clicking through the trees: another pine cone falling, skittering down through layers of branches. But the noise is persistent, nearly rhythmic. I crawl to the edge of the dune leaving a spoor of myself trailing through the reindeer lichen. The shoreline is a waxy yellow colour, bright with gorse flowers. Barefooted, I feel the sharp bite of wood ants objecting to my toes. I roll away from their nest and peer down from the dune. Bloody ants! My feet are stinging from their bites … In amongst the wrack two hooded crows are lifting shellfish a few feet into the air then dropping the shells onto the rocks to crack them open. The crows barely break into flight, more a hoist, a ballet-lift, then hover, release: crack.

Sometimes crows called me out of the trees, their cauldron-croak more noisy than usual. I would go to stand on the edge of the forest and there would be the osprey coming in off the bay again. The crows rising towards him, a buffeting escort.

In poor light above the murky bay I watch the osprey hanging in columns of air. Astonished by how high he is above the water, mesmerised by his flickering hover. I stay with him through the long evening. The tide begins to slip. The patterns in the mud look like the bumps and grooves, the gyri and sulci, of a cortex.

* * *

They planted ten million trees across the dunes to try to stabilise the coast, fixing the sand with a thatch of brash to stop the young trees from being smothered. Still, some were smothered and some trees have half their height buried underground. As the brash rotted it formed a layer of humus, but it is a thin layer and only an inch down my finger finds cool black sand.

One evening I walk through the forest to its northern edge and lie down in the marram grass overlooking the firth. Spread out across the tidal flats are long poles sticking up out of the mud. They look like the remnants of a forest, a ghost forest, overwhelmed by the sea. Gulls perch on the top of the poles like lit bulbs. Later I discover the poles were erected during the war to frustrate enemy gliders from landing in a feared invasion from occupied Norway. I wonder why they have never been removed, but they do not seem out of place in this place of verticals, where a church steeple can suddenly emerge out of the sands, shellfish and pine cones fall and crack, ospreys tilt and dive.

Charles St John, the Victorian collector who shot the osprey near its nest in north-west Sutherland, recalls an incident when he was fishing on a bright summer's evening on the river Findhorn. St John describes hearing a sound *like the rushing of a coming wind.* Yet it was a still evening with barely a breeze and so he shrugs the noise off and continues to fish. A few minutes later he notices that the slack water he is standing in is suddenly sweeping against his feet. He turns to look upstream and sees the terrifying sight of the river rushing towards him *in a perpendicular wall of water, or like a wave of the sea, with a roaring noise, and carrying with it trees with their branches and roots entire, large lumps of unbroken bank, and every kind of mountain debris.* There has been a downpour in the Monadhliath mountains some hours, perhaps days, ago and now the legacy of all that rain is about to hit him. St John has just enough time to grab his fishing gear and run for his life, scrambling up to the safety of a rock above the flood.

That sand can behave like this too, that it can be as destructive as water, that it can flood and ruin like water, is hard to grasp. Yet sand has its own fluidity and can shape and mould objects as water can. Charles St John observed that the stones which the Findhorn rolled down from the mountains to the river's lower reaches were *as rounded and water-worn in their appearance as the shingle on the sea shore.* The trees planted at Culbin which were submerged by the shifting sands emerged again thinner, their trunks tapered where the sand had moved around them. One traveller describes being out in Culbin during a sandstorm, a pressure of weight on his body dragging him down, as if gravity had been increased, his pockets, shoes, clothes, eyes, nose ... filled, saturated with sand. Sand pours from his clothes like water, he is nearly washed away.

Even the mighty river Findhorn is forced by the sand to change its course. One year the mouth of the river is blocked by such a mound of sand the Findhorn has to shift itself to find another way through to the sea. Birch trees, like snow poles marking a buried road, still line the route of the old river. In the forest, pools of water appear each winter along the course of the extinct Findhorn, remembering it, as if the forest almost dreamt the river back to life.

V

Sea Eagle

Morvern Peninsula

The shepherd brings MacGillivray a sea eagle he has shot. The day before, the shepherd had dragged the carcass of a horse up to the spot where, as a boy, MacGillivray tethered his uncle's white hen to a post and lay in wait with his gun. The horse is all ribs and strips of tattered skin. There is a drag-line across the moor where the weight of the horse has dented the heather clouds.

The ribcage perched on the brow of the hill like that looks as if a boat is being built up there. You would not know the hide the shepherd conceals himself in from all the other heaps of stones. And if you glimpsed, from a distance, the shepherd busy about the carcass, charging his gun, then disappearing inside the pile of rocks, you might have thought you'd seen a ghost up there, or spied the entrance to some fairy broch.

It is a fine specimen the shepherd brings him. MacGillivray starts to skin the eagle's head and feet. Its skin is covered in long soft down. *Haliaeetus albicilla*; White-tailed Eagle; Sea Eagle; Cinereous Eagle; Erne; *Iolaire chladaich* (the shore eagle); *Iolaire suile-na-grein* (the eagle with the sunlit eye). MacGillivray gives it the name *An Iular riamhach-bhuidhe-ghlas*, with his idiosyncratic Gaelic spelling, which at times has thrown me off the scent trying to follow him through the Gaelic place-names in his journal. And for several hours I find myself struggling to translate this name, picking over *The Essential Gaelic–English, English–Gaelic Dictionary*, trying to navigate MacGillivray's spelling ... The best I can do with the translation: 'the eagle with the yellow-grey markings', or 'eagle brindled with yellow and grey'. Unless what the shepherd has actually brought MacGillivray is that other species of eagle, that unknown fugitive eagle of the Harris hills, and this was MacGillivray bestowing on the bird its Gaelic name.

* * *

I have begun to lean on William MacGillivray more than I ever thought I would. If I find myself grounded, my journey on hold, held up by work and family and everything else that must come first, if I find myself away from the birds for too long, I've learnt to rely on MacGillivray, his journals, his books, especially his first book, *Descriptions of the Rapacious Birds of Great Britain*, which I carry everywhere, heavy as a brick inside my backpack. And if I become lost, if, as often happens, I cannot for the life of me think how I'm going to find these birds, I've learnt that it's best to revert to MacGillivray, to his writing. And there's always something there – in him – that gets me going again, keeps me heading back out there to search for the birds.

One evening, as I drove home from work, a buzzard slipped from a squat winter hedge in front of my car. I flinched. The bird was huge and close, filling the windscreen, rushing towards me. I was sure I was going to hit it. But at the last minute the buzzard tilted, flipped up, skimmed over the top of the car, cleared the adjacent hedge and dropped down into the field beyond. I see buzzards almost daily along that stretch of road but they had never brushed so close to my car. What would I have done if I had hit the bird? Pulled over, picked it up? But then, what could I have done with its body? I could not just discard a bird so beautiful. And as I drove on I thought of MacGillivray as a boy dumping the golden eagle he had shot on the village dungheap. And only a few years on from that and birds had entered his life, come to possess him, so much so that he cannot relate to his younger self, cannot think what possessed him to discard the eagle. And now the shepherd is bringing MacGillivray this dead sea eagle because he – MacGillivray – has sent out word that he is looking for specimens to study and before he knows it he seems to have become the depository for all the dead raptors in Scotland. He wants to study these specimens because he wants to understand how the birds work and because, above all, he wants to find a new way – a more accurate way – of describing and identifying birds.

All those hours spent sifting through the dead. His desk lit with the bit parts of eagles, oesophagus, stomach, intestines ... For, *it must be obvious*, MacGillivray wrote,

> that a bird is not merely a skin stuck over with feathers, as some persons seem to think it, but an organised being, having various complex organs and faculties, the description of all of which is necessary to complete its history.

So the eagle's stomach is no less important than its bill. And this is what distinguishes MacGillivray's work as an ornithologist, his attention to this detail, his insistence on studying and describing every aspect of the bird. And in this way he advocates a new way of seeing the birds, of looking at them so intently that when he is watching a sea eagle perched like a sentinel on the great sands at Luskentyre, he looks beyond its white-fanned tail, the huge, heavy bill, into the pulse and texture of its form, to draw an extraordinary word map of the bird:

> **Sea Eagle**
> Bill nearly as long as the head, very deep, compressed, straight, with a long curved tip … Stomach large, compressed, oblong, slightly curved, the muscular coat very thin; the two central tendons small and thin … the inner coat soft, without rugæ. Pylorus with a valve formed of three projections. Intestine very slender, nearly uniform in diameter until towards the extremity, when it is considerably dilated … Eyes large, overhung by a thin projecting eyebrow; eyelids edged with bristly feathers … Head broad, rather large; neck rather long and strong; body full and muscular, of great breadth anteriorly; wings long … The cere and bill are pale primrose-yellow; the iris bright yellow; the tarsi and toes gamboge; the claws bluish-black. The general colour of the head, neck, breast, back, and upper wing-coverts, is pale greyish brown, the hind part of the back passing into wood-brown; the belly and legs are chocolate-brown, as are the lower tail-coverts and rump-feathers, some of the upper tail-coverts being white. The primary quills and alula are blackish brown; the base of the primaries and the greater part of the secondaries tinged with ash-grey. The tail is white, but a small portion of its base is deep brown.

I think how lucky I am when a bird such as that roadside buzzard flares suddenly through my day. Those encounters that come out of nowhere, that sear you with their proximity, their unexpected beauty. That sea eagle, for instance, that lived out its life in captivity in the gloom of the old library at Edinburgh University who one evening, as MacGillivray was passing, jolted itself momentarily out of tameness, reached out a leg and clutched MacGillivray's shoulder with its talons. MacGillivray said the eagle did no material damage but I wonder if it was MacGillivray's own aura of passing wildness that made the erne relapse into its wilder self. That wildness that caught up with MacGillivray and shook him periodically, which was not really a wildness, as such, but a fever to be gone, to be elsewhere, out there among the birds.

My journey south was not a fluid thing. There were stops and starts, overwinterings, snatched moments to pick the journey up again. One place that snagged me, kept hoicking me back time and again, was the Morvern peninsula in the West Highlands. So often I returned to walk around its long coast looking for *Iolaire chladaich* (the shore eagle), I realised, unintentionally, I had walked almost the whole way round the peninsula. And then, of course, there were gaps in my circumnavigation which niggled at me, drew me back again to try and complete the circuit.

I went to look for sea eagles in Morvern more out of instinct than any real certainty I would find the birds there. I thought perhaps there might be an overspill of sea eagles from the neighbouring Isle of Mull, which holds the highest breeding density of the birds in Scotland. And really it would have been easier to have gone to Mull, easier to look for the birds there. But, stubbornly, I wanted to try and find the eagles with difficulty, to come across them, if possible, unexpectedly, to immerse myself in the birds' landscape, and wait for them there. I don't especially recommend this approach. Morvern battered me like no other place on this journey. Orkney's luck deserted me, the sea eagles were elusive, the midges voracious. But I got fixated on the idea of trying to find sea eagles there, of walking around Morvern's largely trackless perimeter. I became as stubborn as MacGillivray did on his long walk to London, wanting to complete the journey on foot, despite his hunger and exhaustion, despite the passing stagecoaches which could have scooped him up and whisked him on to London and the British Museum.

But for all the battering my muscles took in Morvern, the homesickness and midge bites, I was glad to be walking in such a remote place, sleeping out in deserted bays, sheltering in the birch and oak woods, waiting for a glimpse of the shore eagle, scanning every tree for the bird's tall grey shape, willing every rock along the shore into flight.

I spent many hours in the steep-sided gullies that carried the burns down from the mountains into Loch Linnhe and the Sound of Mull. Corridors of beauty: primroses the colour of an erne's cere; marsh hair moss and wood hair moss; the slender light of silver birch; grey-green lichen, delicate as filigree, covering every branch and twig like a mist. A cuckoo on a fence post – a hawk mirage – confusing me in my raptor stupor. A buzzard coming in across the sound from Lismore, waking gulls from their roost. When the cliffs rose too sheer, I skirted round them, using the gullies as avenues into the mountains, moving from the shore to the high tops, from sea eagle to golden eagle.

One night, lying in my tent on the loneliest shore in Scotland, I heard footsteps on the shingle, something disturbing the stones, a heavy crunch. Then the sound of pebbles rolling over each other as they fell away down the steep shingle bank. I unzipped the flysheet and stepped out onto the grassy sward. Twelve feet from me stood a red deer hind, long grey neck, ears pricked up, snorting, huffing at me, huge in the drained light. Later, I woke to the sound of the whole shingle bank collapsing around me and stepped out to see a large herd of deer passing my tent in the half-dark, hooves clattering the stones. I watched their black shapes wade into the deep bracken above the wrack line until only their heads and tall necks were visible and the herd floated away up the mountain's lower slope.

One of the things I learnt from my visits to Morvern was how impatient I could be, always wanting to peer around the next headland to look for sea eagles there, fixated on the idea of circumnavigating the peninsula. I was like my teenage self scampering over the moor on Lewis, impatient to show my mother the lochan with its red-throated diver. But I knew I was trying to cover too much ground on Morvern, knew I needed to slow things down and concentrate on looking for sea eagles in a smaller area of the peninsula. So I went back to Morvern one more time to try and steady myself with patience – Gordon's eagle-watching patience – and wait for the sea eagles in a place where I felt they might come to me.

The plan was this: to curb my impatience, to stop myself eating up the distance, I would lock myself into a small uninhabited island in Loch Sunart off Morvern's northern edge. I would wade across to the island at low tide then pull the tide up behind me. The island – Oronsay – would be my lookout post, my home for several days, a place to wait and wait and watch for sea eagles.

A slippery crossing over the rocks between Doirlinn and Torr a' Choilich, like walking over the backs of sleeping seals; a tidal channel, a sea-moat, the seaweed ankle-deep. Then: bluebells, bracken shoots, a fringe of stunted oaks and I am scrambling up the island's rocky eastern slope. Oronsay: what a giddy, youthful feeling, a whole island to myself to explore. But first things first, looking for fresh water and walking all around the island that afternoon and finding only brackish trickles and sluggish pools amongst the reeds, tiny streams with the shortest of lives, done for by the sea before they can get going.

And finding too, that afternoon, crumbled houses with nettle floors and window sockets looking out towards the hills of Ardnamurchan. And one house still with its chimney stack in place, its high gable end, still angled like a house, not yet collapsed, not yet rounded off by years of wind like its neighbours were. How on earth did anyone make a living here in this rocky, brackeny place, the island virtually gnawed to its core by the sea …

That night I pitch my tent above a small bay on Oronsay's northern shore, between an otter's whistle and a cuckoo's call. All night it rains and I listen to the slack burn behind me coming alive beneath the yellow flag irises. The rain on the tent keeps me awake and I stay up reading about Oronsay's ghosts:

> There was yet another eviction on the estate of the late Lady Gordon of Drimnin, and as this was a particularly hard case, which took place only about fifteen years ago, we feel in duty bound to refer to it as showing how completely the Highland crofter is in the power of his landlord, and however unscrupulous the landlord may be in the present circumstances there is no redress. The circumstances are as follows: About forty years ago, when the sheep farming craze was at its height, some families were removed from the townships of Auliston and Carrick on Lady Gordon's estate, as their places were to be added to the adjoining sheep farm. The people were removed to the most barren spot on the whole estate, where there was no road or any possibility of making one. They had to carry all manure and sea-ware on their backs, as the place was so rocky that a horse would be of no use. Notwithstanding all these

> disadvantages, they contrived through time to improve the place very much by draining and reclaiming mossy patches, and by carrying soil to be placed on rocky places where there was no soil. During the twenty-five years they occupied this place their rents were raised twice. Latterly, with the full confidence of their tenure being secure, they built better houses at their own expense, and two or three years afterwards they were turned out of their holdings on the usual six weeks' notice, without a farthing of compensation for land reclaimed.

In the morning the island is drenched and heavy. The bay smells of bog myrtle and rusting kelp. Crawling out of the tent I startle an oystercatcher who peels away across the water, calling loudly. It is cold this morning and I work quickly to make tea and porridge. It has stopped raining but there is still so much moisture in the air my woollen mittens are soon damp and I have to keep wiping the lenses of my binoculars.

I spend the rest of the day perched on Oronsay's headlands, scanning the length of Loch Sunart. North across to Risga, Glenborrodale, Eilean Mòr. East to Carna with its hill still being worked on by the rain. West to Ardmore Point on Mull, to the rocks of Sligneach Mòr and Sligneach Beag. Further west, the outlined hills of Coll. Waiting, scanning every inch of shore.

The account I read last night of the people evicted from Auliston and Carrick, that *particularly hard case*, was recorded by the Napier Commission (the royal commission set up by the government to assess the conditions of crofters and cottars living in the Highlands and Islands) when its commissioners arrived in Morvern in August 1883. Because the place the Morvern crofters referred to in their evidence as *the most barren spot on the whole estate* was unnamed, it took me a while, and some more digging, to realise they were referring to Oronsay, a place *where there was no road or any possibility of making one*, a place *so rocky that a horse would be of no use*. The people from the townships of Auliston and Carrick were removed to Oronsay and then, in turn, removed from the island they had worked so hard to make habitable, picked up and herded on again.

– What became of the people of Oronsay?
– One of those who was in Oronsay was the last delegate, another is in Glasgow, he removed to Glasgow, and two or three are on the adjoining estate of Mr Dalgleish, Ardnamurchan.
– The club farm was abolished, and the people had to go?
– Yes.
– Who has it now?
– A large farmer.
– What is his name?
– Donald McMaster.

...

– As I understand your statement, the people were removed for the benefit of the sheep farm, and you may say for the benefit of the estate?

– And for the benefit of themselves.
– But the people were not made the judges of their own benefit?
– They were not asked in the first place.
– What I want to arrive at is this, the people were virtually and substantially removed for the benefit of the estate, in order that this sheep farm, or some other part of the estate, might be more profitably administered and held; in removing the people did the proprietor, in consideration of their number and poverty, and the difficulty of obtaining other places, make them any allowance or gratuity?
– Not to my knowledge.

And here is the rain again, coming down off Beinn Bhuidhe from the south, crossing Loch na Droma Buidhe, then rushing at me as I crouch, hunker down on Oronsay's headlands. All I can do is turn my back and wait for the rain to clatter over me.

– Is there any use in beating about the bush; is it not the fact that those people were removed solely and entirely because they were in the way of sheep?

Some of the things I see that day: cormorant, greylag, skua, tern, black-backed gull, ringed plover, great spotted woodpecker, sandpiper, raven, pipit, hooded crow, heron after heron after heron beating low across the loch. And do not see: the buzzard I hear calling in the woods above Doirlinn, the pine marten whose scat I find lying on a mossy boulder, the erne, the sea eagle, the eagle of the shore.

What sort of bird is it that taps MacGillivray on the shoulder in the dusk of the old library in Edinburgh? A robber baron, a pirate, a sluggish vulture of lakes and fjords ... Much of the time the sea eagle is perched – cormorant-like – on a rock or tree beside the shore. Except, not like a cormorant at all, because if you see the eagle's great bulk on an islet in the loch it looks as if a person is standing there, not a bird at all. Their sheer size: nothing prepares you for that.

Diet: anything, really ... Fish snatched from the shallows, a low glide above the water, pause, a brief hover ... then talons taste the water, seize and grip the fish and the sea eagle is beating its great wings, heaving itself into height. For such a large bird the agility with which the sea eagle plucks fish from just below the surface is astonishing. Wing tips might brush the water as it struggles to rise, but the sea eagle does not immerse itself in the water as the osprey does, there is none of the splashing crash of the osprey as it plunges after fish.

Shetland fishermen used to anoint their bait with sea-eagle fat believing this would bring them luck, such was the great bird's fishing prowess. And just as bait smeared in sea-eagle fat could make all the difference to the day's catch – and a dream of home was all it took to haul in nets and head for home – so the fishermen of the Northern Isles had to tread carefully around language, using their own sea language when they were at the *haaf* (the fishing). Everyday things – pigs, rabbits, fish, the minister – had to be skirted round, ducked under, referred to aslant. So the fishermen could not, for instance, call the birds they saw from their boats by their real names, as to do so would be to tempt ill luck. Cormorant, puffin, sea eagle … all names that were taboo and could only be spoken of in code like a distorted echo of each bird's call. A defiant language of the sea, of kennings and circumlocutions, rooted in Old Norse, holding out against the tide of language from the south. And when they spotted a sea eagle soaring above their boats bobbing in the thick swell off the cliffs of Hoy and the cliffs of Fetlar and Noss, the fishermen would refer to the bird in their strange sea language, calling the sea eagle *Adnin* or *Clicksie* or (my favourite) the *Anyonyou*.

Besides fish, the sea eagle will take shore birds, moulting geese, injured wildfowl, young kittiwakes. A hunting sea eagle will spark the shore birds into the safety of flight, unlike a passing peregrine, whose presence grounds everything. Also: hares, rabbits, occasionally a hauled-up sleeping seal. Sea eagles will pursue gulls – even ospreys – to make them spill their catch. MacGillivray wrote they were especially fond of dogs. And of course carrion: sea eagles have a great propensity for carrion and the birds (described in the chronicles as the 'grey-coated eagle') were often observed cleaning up the aftermath of Anglo-Saxon battlefields, rehoming the souls of the dead.

Though not so vulturine, not so sluggish as we tend to dismiss it for being. Male sea eagles, when providing for their young, have been seen to almost kill at will, swooping repeatedly at diving birds, harrying ducks, cormorants, auks, tracking an eider drake's bright tracer under water as the eider keeps on panic-diving, plucking it from off the water when the duck is too exhausted to dive again.

MacGillivray came at the sea eagle, of course, through observing its anatomy and behaviour, showing how the sea eagle is a convergence of other birds – gull, skua, vulture, osprey – tracing the way these birds gradually form a passage into each other. And in his own way MacGillivray was anticipating Darwin, observing in his studies of birds of prey how the different species had grown alike, evolved to equip their rapacious lives, whilst their anatomies retained the blueprint of their separate origins.

So, part skua, part vulture, part osprey. Above all, perhaps, a cousin of the osprey, in the way the sea eagle can pluck a fish from the shallows, the way that horny spicules line the soles of its feet to help it grip the fish, the way their Gaelic names intermingle – *Iolaire uisge* (eagle of the water) for the osprey/*Iolaire chladaich* (eagle of the shore) for the sea eagle. And above all like an osprey in the way their stories – their melancholy histories – converge. The sea eagle driven out from the great English river estuaries, from its refuges on Lundy, Shiant and Hirta, from the Lakeland crags and the cliffs of Jura. So by 1914, when Gordon is writing his chapters on the osprey and sea eagle for his book *Hill Birds of Scotland*, he is drafting both birds' obituaries. The last reported nesting of a sea eagle occurs on Skye in 1916. And then there is one, a lone female sea eagle – an albino bird – haunting the cliffs of North Roe in Shetland.

When it comes, the end is a hurried thing. The shepherd on Harris tells MacGillivray that he believes the sea eagle to be far more common than the Black (the Golden) eagle. So it's as if the sea eagle falls through a plughole in the sky, to go from being so numerous (in some districts more numerous than the golden eagle) to that lone Shetland female bird in the space of seventy years. It seems that the sea eagle took the greatest share of blame as a lamb killer (more so than the golden eagle) from the new breed of shepherds, jealous of their flocks, that took over the townships of Oronsay, Sornagan, Carraig and Auliston and all the green lands of Morvern and beyond.

The sea eagle, too, proved an easier scapegoat – more easily reached along the shores – than the golden eagle with its skulking ways deep in the mountains. Carrion laced with strychnine was a favoured method for getting rid of the birds. Burning peats were lowered down the cliff to set fire to nests. Failing these, the shepherd would put out some bait then fold himself into a hide of stones to lie in wait with his gun.

And a dead sea eagle fetched a good bounty. In Orkney, a hen from each house in the grateful parish was paid to the person who killed an erne. Ten shillings was the going rate on Skye. The bird's parts had their uses too: apart from the fat which fishermen coveted, the sea eagle's long broad wings would serve as a useful broom. On cleaning days, peering through an open doorway, women could be seen moving back and forth like injured birds, dragging a great broken wing across the room.

28 June 1914, North Roe, Shetland

The artist George Lodge is walking across the hills of North Roe with two companions and the watcher James Hay. It is a day of wind and wet mist that drives across the hill. They struggle to keep up with Hay, who moves like an antelope across the tussocks and deep moor grass. The party is making for the Red Banks where the last resident sea eagle in the British Isles can be found. Hay tells them that she is an albino specimen and has lived here going on thirty years. For a long time she was paired with a mate, but they bred for the last time in 1908 and shortly afterwards the male departed. Since then the female has lived out here on these cliffs alone.

They see her from a long way off, a white spot on the rocks below the old nest. As they approach she flies off and out across the bay and they do not see her again for the rest of that day. She looks as white as a gull when flying. Lodge makes an oil sketch of the nesting cliff. He works for an hour and a half but the weather is awful and after a while he is too cold and cramped to continue, so he packs up his easel and heads homeward.

30 June 1914

Very windy today. The party heads back out to the Red Banks in search of the sea eagle. She is nowhere in sight when they get there and it is too windy to paint, so Lodge makes a pencil sketch instead. The nest looks like a mass of rubbish and is hard to distinguish against the backdrop of rocks. He notices a hooded crow mobbing something round the corner of the cliff. His friends go to investigate and out flies the eagle pursued by several hoodies. He has a good view of her as she passes in front of him, notices that her primaries are not white but appear to be a light brown.

10 July 1922

Eight years later. Lodge is back on Shetland and making for the Red Banks to look for the albino sea eagle. Hay tells him that she has not been seen since 1918. On the way to the old nest site the route is blocked by a burn in spate and Lodge has to trek for two miles before he can cross it. Finally he finds a place to ford the burn and heads back downstream until he reaches the cliffs where he made the sketches in 1914. There is now not a trace of the old nest.

There is a rumour on Shetland that the eagle was shot in 1918, though no one knows what happened to the body. Lodge crouches down out of the wind and scans the cliff face. He cups his hands either side of his mouth and calls out to the cliffs, listens to the pause, the ricochet, the dissipating distorted echo:

Anyone there
Anyonyou
Anyonyou

After the sea eagle has gone the land feels emptier. All those place-names in the evicted landscape. *Earnley* (the wood of the erne); *Arncliff* (erne's cliff); *Creag na h-Iolaire* (eagle's rock); *Cnoc na h-Iolaire* (eagle's hill). And after that lone white female is shot in Shetland in 1918 there are decades of emptiness. But then, in parallel with the osprey, the sea eagle starts to re-emerge into some of these emptied spaces. Young birds are brought over from Norway in the mid-1970s and carefully, painstakingly, reintroduced to Scotland's coasts. Their numbers are still small and like all large birds of prey they are slow breeders (slow to reach breeding age, slow to rear their young and rarely raise more than two chicks per year). But sea eagles have returned and they are gradually spreading along these shores, reanimating the rocks and hills and woods that were named after them.

Besides my visits to Morvern I made trips to other places to look for sea eagles. Recently the birds have been reintroduced to Scotland's east coast and I went to seek them there along the Tay estuary and made several midwinter visits to Loch Leven, where the eagles are sometimes drawn in cold weather by the large flocks of wildfowl that gather on the loch. Occasionally I caught distant glimpses of the eagles but always they were a long way off, crop-filled, static, sitting on a rock or at the top of a distant tree. It bothered me, my lack of any really good time spent watching the birds. I wanted very much to observe how they flew, to study their shape in flight, to watch them hunting. I had to try one more time. So I went back again to the West Highlands towards the end of June and this time focused on a small area further up the coast from Morvern where I spent several days watching the shoreline for sea eagles. I was so glad that I went back.

Early morning, looking down from the cliff at the narrow sound below. The tide is rising and the narrows are a frantic squall of whirlpools and standing waves. It looks like the sea is coming to the boil. A huge volume of water is being squeezed through the channel and the narrows bubble over under the pressure. The air is a cauldron of gulls, swirling above the rushing tide, feeding on the fish being pushed with the tide through the sound. Blackbacks, herring gulls, common gulls, cormorants, the white wing-flash of a great skua, a lone gannet … There are dozens of seals too feeding amongst the gulls, snorting, slapping the water with their tails. The morning is so still I hear the seals huffing and whacking the surface long before I reach the cliff. And the noise of the seals is so loud at first I think it is being made by much larger beasts, as if a pod of whales were surfacing there. Some of the seals drift very close, slow black shapes through the clear water beneath the cliff. There are so many seals I could hop from one to the other like stepping stones across the narrows. In places where the tide is running fastest there are seal–gull skirmishes, thefts and squabbles in the thrashing water.

A slight rise in the noise from the gulls, like an increase in pitch of their agitation, and there is the sea eagle flying up the channel. He is following the shoreline, slow shallow wingbeats, but moving quickly. Brown and grey like a red deer's winter coat. Long-fingered primaries fray his wing tips. The most striking thing about his flight silhouette is his long thick neck and the neck is elongated by his huge yellow bill. In flight he is not as outrageously large, nor as cumbersome, as I had expected. There is speed and agility there and he twists easily through the swirling cloud of gulls. His short tail seems to check his size in flight, it gives his shape a squatness (much as a buzzard's tail does). So what you see is mostly wing and neck (unlike a hawk in flight where what you notice is the long narrow tail). It is when he is shadowing a herring gull, or being mobbed by a hooded crow, that the sea eagle comes into his scale. He swamps these smaller birds with his size and the great width of his vast, broad wings sinks home.

I am overjoyed to see him. After all those days spent walking around the coast of Morvern, sleeping out with Oronsay's ghosts ... A chance, at last, to spend some time watching – properly observing – a sea eagle. The conditions are perfect: clear and bright and the eagle appears to want to linger here, where the fishing is good. I have found a seat above the cliff, a mossy cushion amongst the heather where I can look down on the whole sweep of the narrows and scan the woods and rocks on either side.

The gulls have brought the eagle here. And it is the gulls that announce his arrival through the change in pitch of their squabbling. If I cannot locate the eagle I learn, over the coming days, to wait and listen out for the gulls' signal that he is approaching. It is the gathering of the gulls as the tide starts to swell and push through the narrows that ushers in the sea eagle. He is here because he is an opportunist, a fisherman, a scavenger; he is here because of the gulls. The way the eagle interacts with the gulls – the dynamics of this relationship – is fascinating to watch and I am quickly caught up in the drama of it.

Most of the time the eagle is perched near the top of a young larch tree overlooking the sound. He sits up there and I sit opposite him on the other side of the narrows. What is most conspicuous when he is perched like this is his dun grey head and neck. I can also see his heavy yellow bill and there is a flash of yellow a little further down where his talons lock around a branch of the tree. If I lose sight of him – if I have to rest my binoculars to clean the lenses or if a seal-slap below the cliff distracts me – I can usually pick the eagle up again on his perch by scanning the trees for the pale yellowy light his head and bill emit. I can see now why MacGillivray gave the sea eagle that peculiar Gaelic name, *An Iular riamhach-bhuidhe-ghlas* (the eagle brindled with yellow and grey). Those are the bird's colours (the mature adult bird's colours) precisely. He sits up there on the tree, very tall, his head and neck glowing faintly like a dim harbour light. His long wings folded and hanging down behind him make him taller, lengthen his back. The top of the young tree is bent under his weight.

Then he is up and has latched on to a gull. When he leaves his perch there is nothing awkward in the way he detaches his great bulk from the tree. He just steps off, huge wings outspread, launched at once into a low, fast glide across the water. Why does he select this particular gull? What is it about this gull that makes it stand out from all the others churning and squabbling over the narrows? There must be something that gives this gull away, a weakness, a vulnerability, a visible fish-bulge in its crop, something that signals to the eagle that it is time to leave its lookout perch and give chase.

What follows is a shadow dance. The eagle latches on to the gull and proceeds to follow it everywhere. His victim is an adult herring gull, a large bird in itself, but utterly dwarfed by the sea eagle's heavy presence. The eagle does not attempt to attack the gull, simply looms over it like a dark cloud, stalks the gull for as long as it takes, staring down the gull, until the gull can take no more of this thing that is so large and overbearing that its wings block out the sun, locking the gull in a permanent shade.

When the gull finally relents I see it drop something, a tiny morsel of bright fish. And in an instant the sea eagle has shifted into something altogether else. It defies its bulk, its size, its ponderous, sluggish reputation and transforms itself into a bird I did not anticipate. The eagle twists into a burst of speed and swoops down after the gull's spilled catch, but when the gull drops the fish, eagle and gull are too low above the water, and despite his sudden rush of speed there is not enough time for the eagle to intercept the fish. He sees that he will not reach it in time and already, before the fish has even hit the water, the eagle has given up, pulled out of his dive, and is beating back towards his pine-tree perch.

When I check again the eagle is up and flying across the narrows towards me. I sit very still watching him loom and grow through the binoculars. I have a fine view of his long neck and pale nape. He keeps coming straight towards me. I wasn't expecting this ... what is he doing coming so close? Surely he has seen me? At the last minute – whoosh – he banks upwards, above my head and lands on a rock on the hillside behind me. His head and the shouldered tops of his folded wings are silhouetted by the backdrop of blue sky above the hill. Straight away a hooded crow is swooping down at the eagle's head, dive-bombing him repeatedly. Each time the hoodie swoops at him the eagle flinches, recoiling his long neck from the attack. It is the raptor's curse, this relentless mobbing; gulls and corvids, especially, are the raptor's bane. Mobbing is their way of saying: *We can see you, we don't fear you, we don't want you here ...* The hoodie will not leave the eagle alone. The crow keeps on haranguing him until the eagle has had enough. And he is up again, flying back across the sound to find a quieter perch.

The next time I see him he is harassing a gull, breathing down its neck, following the gull's every twist and turn. The gull does not show any hint of panic. It is unhurried by the eagle's pursuit and does not try to shake the eagle off. They reach a sort of equilibrium, a resignation that the dance just needs to be seen through to the end when the gull must jettison its catch if it's to get the weight of eagle off its back. So they go like this for five minutes until, at last, the gull spills its catch, and this time I see it is a large fish and I follow it as it falls through the air sparkling in the sun and watch it hit the water with a splash. The eagle flexes into speed and shoots after the falling fish but once again the drop-gap is too shallow and he does not have time to catch the fish before it hits the water. Once the fish is in the water the eagle makes no attempt to retrieve it.

The gulls are still here in numbers, working the narrows. The eagle sits on his pine perch watching them. When he is up amongst the gulls I notice how clean and bright his white tail is. It is very noticeable, this flash of white across the tail, whiter than the gulls themselves. Seven herons come croaking up the sound. After them, flying in the opposite direction, a small flock of guillemots, in tight formation, skimming the waves. When I turn back from the guillemots to the shore the eagle has gone from his perch. I look up and see he is climbing high above the sound. Two thousand, 2,500 feet … What is he doing up there so far from the gulls and the fishing? Then I see there is a gull up there with him, trying to outclimb him. The gull is just a speck. The eagle looks like the gull's swollen shadow. For ten minutes they turn about each other in the high air. I keep my binoculars fixed on them waiting for the moment – there it is! – the fish is dropped and this time they are so high up and the fish has so far to fall, the eagle has all the time in the world to flick down in a sudden charge of speed and pluck the fish from the sky.

In the slack tide of the afternoon I walk over the hill behind the cliff. Taking my time, trying to refocus the scale of my gaze from the eagle to the small and near-at-hand. Meadow pipits and grasshopper warblers are out on the hill. Everywhere there is the rapid trilling sound of the warblers and the speckled breasts of the pipits. I meet a dragonfly on the path, as long as my hand, clinging to a thick blade of moor grass. The dragonfly is motionless, undeterred by my shadow. It does not even flicker when I kneel down beside it and place my camera just an inch away to photograph the bright yellow-black traffic-signal stripes along its abdomen.

On the other side of the hill I reach a deer-fenced forestry plantation. A pair of buzzards are calling and swirling above the trees. I follow the buzzards into the wood and find a track that starts to drop steeply down towards the sea. At the high point of the track I pause to scan the canopy for the buzzards again when a huge bird – nothing like a buzzard – flies across the clearing right in front of me. It is the sea eagle, circling just above me and so close that, if I were to climb this stumpy pine, I could reach up and feel the downdraught from his wingbeat. In a quarter of a mile I have gone from dragonfly to meadow pipit to buzzard to sea eagle, as if I were walking through layers of magnification.

I stand very still in the middle of the track and the eagle circles above me, low over the trees. He comes around in a wide arc and as he turns to face me he suddenly flinches and veers to the side, as if swerving to avoid me. He banks awkwardly into a different angle of flight. His huge wings push down under him and the wing tips almost touch each other beneath his body as he pushes and hauls himself away from me.

I saw the moment when he saw me. He was close enough for me to see his left eye as it clocked me standing in the middle of the track. In that instant, I saw how the shock of my shape triggered something in him and caused the sea eagle to flinch and reel away.

In the early evening I am back above the cliff to catch the rising tide. The gulls and seals are there working the narrows. I have not been there long when I hear that sudden shift in tempo from the scolding gulls, and there is the sea eagle beating up the channel. This time he is flying very low above the water, much lower than I had seen him in the morning. What strikes me, too, is the speed and line of his flight: he is heading straight for something. He rises a foot or two, still flying in the same direction, accelerates, then drops down again towards the water. His wings pull back behind him as if he is about to land, both feet push out in front of him and his talons brush the water scooping out a glint of fish. Then he is pulling up and away from the sea and this time gulls are mobbing *him*, rollicking around him, trying to tip the eagle from his catch.

I keep my binoculars fixed on the eagle. The gulls are frantic around him, the noise of their calling has ratcheted up another level. They coil around the eagle as if trying to bind him in a giant knot. But the eagle seems oblivious, he sticks to his course, beating steadily away across the sound, and the gulls begin to dwindle in his wake. Once he has shaken off the gulls the eagle begins to circle over the lower slopes of the mountain. It looks as if he is hunting there, foraging over the ground, and I am puzzled: he has his fish, why is he not returning with it to the nest? Then I realise he is not foraging, he is climbing. He has been looking for a thermal to take him up the side of the mountain and very slowly he is spinning upwards, gaining height. If I drop my fix on him whilst he is flying against the mountain's backdrop I find it hard to pick him up again, the mountain's browns and greys fold the eagle into it. But once he clears the summit and his shape is set against the sky, he is the only black dot up there for miles and I can follow him easily as he keeps on rising. At around 4,000 feet he suddenly turns and shoots away to the south in a steady downwards glide. He has chosen the path of least resistance: once he has gained sufficient height it is a downhill glide and he can cover the distance back to the nest (which must be several miles away) in a matter of seconds.

What a wonderful thing to witness, the sea eagle taking a fish from the water like that. I thought the eagle would revert to harassing the gulls as he had during the morning shift. And it was as if he had spied that fish from a long way out, the way he stuck to such a straight course flying fast and low up the narrows. Here was a bird that kept dismantling my preconceptions of it. In an instant the sea eagle was capable of shifting into astonishing poise and speed and skill.

I am back on the cliff early the next morning. I do not want to miss a beat and I spend most of the rest of the day watching from my mossy seat above the cliff. In the late morning a large school of dolphins comes into the bay, gulls sprinkled above them, drawn by the fish the dolphins are herding. There are so many dolphins, the water churns around them as they travel, and there is such a ferment of gulls following the pod, it looks like a weather front is moving up the channel.

Much of the time the eagle is perched on the shore, or more often at the top of the larch tree. Each time he returns to his perch he approaches the tree from very low down, at the last minute swooping up to the top of the tree, wings labouring, hauling him up the last few feet. He makes several long-distance glides from the tree low across the water, as if he has spotted something just below the surface from a long way out. Most of these flights are aborted and he heads straight back to his perch. Occasionally he lands back on the tree facing the wrong way and has to hop himself around to face the water. When he does this the branch he is sitting on shakes beneath him. Before he settles into his perch he pulls his tail up behind him and it flashes white, once, twice across the sound towards me.

On one occasion he catches something close to the shore, though he is too far away for me to see what it is. Perched on the shore he is camouflaged, his brown body blending into the wrack line of rusty kelp, his pale grey head becoming just another rock. It is hard to keep track of him when he is on the shore like this. I think I have him, refocus the binoculars, then see that all the time I have been staring at a rock ... It is at times like this that he *is* his Gaelic name, *Iolaire chladaich* (the shore eagle).

Sometimes a solitary gull finds the eagle at his perch and there follows an endless round of mobbing. If the gull comes too close (which it doesn't often) the eagle raises his bill towards the gull, extends his long neck to jab it away. But the gull does not relent. It's as if it is locked into a flight path, swooping down at the eagle, persistently, rhythmically, loop after loop, in a shallow curve above the eagle's head.

VI

Goshawk

Scotland–England Border

Here is MacGillivray approaching the Border, walking into autumn through Kirkcudbrightshire, hair a little sun-bleached, shoulders calloused from his knapsack's weight. Muddy splash marks rise up his coat, a crescent pattern to the mud spots like an offshore reef.

It is almost a month now since MacGillivray left Aberdeen on his higgledy walk to London and he cannot put off leaving Scotland any longer. I have (can I say this?) grown envious of his feet, he never seems to complain about them. But are they really immune to all that pummelling: Aberdeen to Fort William, 161 miles; Fort William to Glasgow, 134 miles …? Seldom more than two meals a day, though he sometimes sounds as if he would like to live off less:

> Bread and water will do very well for the greater part of my journey, for many a better man has lived a longer period than will be allotted to it, on worse fare …

But his poor feet, surely they are beginning to ache? In Morvern my feet turned a strange crinkled yellow, the skin softened like margarine (how I wished I had followed MacGillivray's recipe and bathed them in whisky and liniment of soap …). The only time I hear MacGillivray grumble is when he is bitten by bed bugs in Glasgow. He burns up in a fever, can barely open his left eye, and a tumour swells on his cheek to the size of a hen's egg. How he wishes he could have exchanged that bed in Glasgow for a couch of grass on the side of Cairngorm … And now he is so wary of the blighters he sniffs the air of every inn for the bugs' peculiar coriander scent, combs his mangy rooms with a pin, prising the bugs from their daylight hiding places, the seams and chinks of bedposts, the gaps beneath the skirting board, till his pin holds its own length in wriggling bugs, like a keeper's gibbet or the shrike's thorn.

I managed little better with bugs in Morvern, beating my own world record for parasitic ticks, each night lying in the tent, inspecting my skin by torchlight, extracting the ticks with tweezers one by one. Once, in desperation, I tried smoking a pipe to ward off midges, but smoking it was eye-watering misery: I could not get any draw on the tobacco and kept having to relight it, coax and suck it back into life like a damp bonfire. MacGillivray recommended decocting tobacco for the treatment of bed bugs. Either tobacco, or Irish lime sprinkled across the floor. Once, staying at a London inn, he was so tormented by bugs biting his face, neck, arms, back and legs, even the crown of his head, that he resorted to using candle tallow to rub against his skin where he had been bitten.

Glasgow to Portpatrick, 99 miles; Portpatrick to the Border, 107. Expenditures: half the ten pounds he set out with from Aberdeen, the remaining five tucked in a small pocket in the inner side of his flannel vest, though he is certain he won't need all of it. London from the Border at a steady trot, twenty shillings should cover it. *Bread and water* (repeat the mantra!): *better men have lived on worse.*

Five hundred and one miles clocked since he left Aberdeen a month ago. Bearing up pretty well, the swelling from the bug bites going down, a touch of homesickness now and then (he had it in Fort William when he came into the orbit of the Isles). But doing well, considering: on budget, feet enviably robust, approaching the Border, heading south.

* * *

Early morning above the forest, a smoky winter light. The wind starting up, stirring rooks and jackdaws into plumes of soot. A sparrowhawk, skimming the canopy, rises to meet a woodpigeon. No intent to the flight, it just goes up to have a look. The pigeon flinches and the hawk dives back towards the trees. Then a pair of ravens are up, clearing their throats, a glint of brightness, like polished mahogany, in their thick beaks. The ravens sound different here, their voices dampened by the trees. Nothing like the raven in the Outer Hebrides, jostling with the golden eagles, its call cleaned and sharpened by the cliffs, carried down the glen by loch and gneiss.

Everything about this morning feels slow and hushed. The muted ravens, the mushy light, even the sparrowhawk's pigeon-jink had none of the hawk's usual quick-fire dash. There are mornings like this when the forest feels cold-blooded and everything in it seems to take an age to limber up.

Then crows are crossing the forest in twos and threes, ambling, tugging at each other. A buzzard circling to the right of me. A goshawk displaying a long way off. The goshawk is too far out and I cannot get any feel for the bird, any purchase on it. Other than I notice it is roughly the same size as the buzzard, though not as dark and not as compact, not as rounded in its shape. Occasionally I pick up flashes of white showing on the goshawk's plumage. The buzzard is circling, rising through figures of eight. The goshawk is flying slowly a few feet above the treetops. A slight, sudden rise as if the hawk meets an updraught and just as quickly it is dropping out of the rise and landing on a tall pine tree.

Through my binoculars I flick between the perched goshawk and a felled area of the forest about a mile away. The clearing saddles a ridge, a whale's back rising above the shallow valley. I keep drawing back to the clearing, studying it, thinking ... if I head up there, that would be a good place to be, that would be a good place to watch for goshawks. Up there, I might get closer to the birds as they find the lift of breeze above the ridge. So I take a compass bearing, drop down into the forest, and make towards the clearing.

Saturday 2 October, 1819. Three miles from Newton Stewart, MacGillivray comes across an outlier from the south, a specimen of trailing tormentil, on the edge of its range, huddled beneath a hedge. He has never seen the plant before in the north and it dawns on him that he is running out of Scotland. He is aware too of how, daily, the light is leaving the land, in the jaundiced grass out on the heath, in the withered bracken. Very few plants are in flower now and sometimes it feels, as MacGillivray approaches the Border, that he is walking in step with autumn as it moves through the land from north to south, snuffing out each flower in turn along his route. He passes through a field of wild carrot, scarlet pimpernel and corn woundwort and picks some of the carrots to crunch as he walks. When he brings the carrot roots up to his mouth it looks, for a moment, as if he has grown a pale yellow beak like the juvenile choughs he saw feeding amongst cattle dung on Barra.

His pockets are stuffed with all the specimens of plants he has gleaned along the way, all his curious digressions. There hasn't been time to press them all and so many plants are lodged in the nooks and crannies of his clothing it looks like they are sprouting from him. The man on horseback MacGillivray passed in the dusk outside Castle Douglas must have thought he had been overtaken by a scarecrow as he watched the figure stumbling ahead of him, wisps of foliage trailing from his cuffs. He is still uncertain how to answer people who stop to ask him about the plants. On one occasion, walking along with a bunch of cuttings in his hand, a woman stopped and peered at him:

– Is that yerbs?
– Yes.
– Ow! That's rushes.
– Yes, it is a kind of rush (It was Hard Rush).
– Fat's this?
– I don't know the English name of it.
– Are you gaun ti' drink affin't?
– No I don't intend to put it to any medical use …

MacGillivray wrote that he left Scotland on his walk to London *without regret*, that Scotland was *too wide a word* for him, that it was the Isles and mountains that really claimed him. But I wonder, did he not feel less sure-footed as he stepped from Dumfriesshire into

Cumberland, did that specimen of trailing tormentil not remind him that much of what lay ahead was unfamiliar to him? And besides, once MacGillivray crossed into England he pelted through it at such a rate, head down, a beeline for London, none of the meandering diversions that Scotland steered in him.

All that walking in circles around the Morvern peninsula, I needed to straighten out. So, when I reached the border between Scotland and England, I decided to follow it and walk along a stretch of its line in the deep goshawk forests above Kielder. I spent the days following broken dykes and firebreaks through the trees, crisscrossing burns and cleughs and sikes. Each firebreak in the forest felt like an avenue for goshawks, every tree along every ride and clearing I scanned for that heart-thumping goshawk meeting. For several days I hesitated over the Border like this, dipping my toes in and out of England. Sometimes I found myself zigzagging, leapfrogging back and forth across the line for mile after mile. In the evening returning to my tent pitched on the boundary, my head dreaming in Scotland, my legs hanging over the Border, pointing south.

Waking on the hillside, and being woken by a sound like a stream gurgling beneath me: red grouse, starting up the day, calling through the heather, their calls burbling under my ear. Fog this morning and the forest has stepped off into whiteness. I go in search of the trees, walking across sphagnum humps, tussocky reed grass, the fog condensing on the sharp reed blades and spiders' webs. In a boggy alley on the side of the fell I find a bloom of grouse feathers, neatly plucked, lying in the heather. I search for the carcass but find nothing. Goshawks are fastidious about preparing their prey, can spend over an hour plucking a pigeon, feathers growing in a ring around the hawk as it works. Afterwards the hawk can feed for an hour, sometimes two. The pile of grouse feathers looked like raptor's work, too neat for a fox.

I fumble through the mist and finally reach the trees. Along the forest's rim I pass through different belts of roe-deer droppings, some fresh from the night before, wet, shiny black, others much older-looking, parched and crumbling. Inside the trees and the forest is dripping with fog-dew. It is dark among the pines with a sense of heavier darkness just out of reach. Beyond the trees there is a brightness inside the mist which seeps a little way into the trees then quickly dissolves in the gloom. Rackways, where thinnings have been removed, wander into the forest then peter out. Deer paths blend into the rackways then veer away again like branch lines or capillaries. I follow one deer track deeper into the forest. Something is glowing up ahead through the trees and I wind towards it like a moth. It is some sort of relic of the Border line, a strange arrangement of standing stones and cairns, gnomish monuments, draped in a moleskin of green lichen that glows almost fluorescent amongst the pines.

* * *

Planting began at Kielder in 1926 and it soon became the largest man-made forest in the country. Sitka and Norway spruce, the Sitka darker, bluer. Scots pine and Lodgepole pine, a redness, like lava, in the fissures of the old Scots pine bark. Bright clumps of larches amongst the pines. Few native trees remain: sallow willow in the valleys, birch in the narrow cleughs, alders along the river banks and on the river's islands of silt.

Wind can havoc a plantation like this. Shallow roots on wet, clay-like soil made the trees especially vulnerable. Early thinning in the forest proved disastrous, letting the winds in to roam through the blocks of trees, yanking at their weak roots, toppling the pines like dominoes. You see the same phenomenon – the same vulnerability – in goshawk feathers. If a hawk becomes stressed through lack of food or, as is sometimes the case in birds trained for falconry, through psychological stress, the bird's growing feathers record the period of distress as a pale line, a band of weakness across the vane. These lines are known as 'fret marks', records of distress, as when a tree logs a period of drought-stress through a narrowing of its growth rings. Feathers with stress marks are vulnerable to breaking along these weakened points. As with the thinning of the forestry plantations, once one feather breaks there is less support for the remaining weakened feathers and these can swiftly follow suit, impeding the hawk's survival, its ability to fly and hunt effectively.

* * *

I keep stumbling across things buried beneath the forest: steadings, sheep fanks, stells ... There is a sense of a busy landscape being drowned under the trees as rapidly as Culbin was smothered by the sandstorm. A few miles below me is the great reservoir of Kielder Water and under its waters sleep all the drowned hill farms of the Upper Tyne.

Every ten years they partly drain the reservoir at Kielder for maintenance. I imagined a dripping world of sunken boats, field boundaries, cavernous farmhouses slowly revealing themselves as the waters dropped. I envisaged folk going down to the shore to gaze at this lost world. And I was sure this is what happened; I had spoken to someone in the village who told me about a road across the bottom of the reservoir, about going to salvage his boat that had sunk. I had read a little about the community of the Upper North Tyne before the valley was flooded, read some of the moving evidence the community had put forward in objection to the reservoir scheme:

> We cannot appreciate why we should be compelled to sacrifice our homes and the whole community spirit of the Upper North Tyne, when equal quantities of water could be obtained from reservoir sites built in areas where there would not be one fraction of the disturbance likely to be caused in this area ...

But the reality is there is no lost world beneath the reservoir. Nothing is revealed when they start to drain it, nor is it drained completely. Every stone of every farmhouse, every tree trunk, was removed when the reservoir was constructed in the 1970s. What was there before the valley flooded – the community of hill farms – was entirely erased. I had just assumed that the remnants of that community would still be there beneath the reservoir. It took a call to the Northumbrian Water Authority to set me right. They told me about the careful disbandment of the farms and houses, the way every tree was uprooted to avoid them later floating to the surface. I wanted to believe that there were still relics down there, signs of what had been drowned in the valley, that the tops of buildings might briefly emerge during a drought, just as they poked up through the shifting sands of Culbin. Instead, every decade, when the waters are lowered, all that is revealed is a vast absence.

Mid-morning and the fog is beginning to peel. Where are you, goshawk? What am I searching for? A chimera, a rumour? A space inside a fir tree where light seeps through, a patch of silvery lichen across a branch ... There are so many things to deceive and delude you when you are searching for – aching to see – the birds. How many buzzards I must have willed into goshawks! How many times I have tried to conjure a hawk out of absence. Even the discarded beer cans I spotted from a distance lying beside the Bells Burn, just over the Scottish side of the Border, were grey and hawk-shaped enough to briefly glint in my imagination, until I was close enough to tell them for what they were and could see how their tin had weathered to a murky silver.

Behind the beer cans, on the opposite bank of the burn, stood a large ring of lichen-crusted stone: a 'stell', a circular walled sheep enclosure, 5 to 5½ feet tall, 30 feet in diameter, a purpose-built snow shelter where flocks could seek refuge from the drifts. I rested inside it for a while, brewed tea, cleaned the glass on my binoculars. The stell was a beautiful construction, remarkably intact. Its wall had been built (as all the walls along the Border are) with two walls leaning towards each other, the gap between them stacked with the *chatter* of smaller stones, both walls then bound together with longer *thruft* stones that reach through the *chatter* like ribs. Built, as well, to be flexible, to bend and sag as the soft ground gives way beneath the wall's weight.

Tree pipits are squabbling amongst the timber brash beside the stell, their sandstone breasts streaked with black rain. The birds bounce from stump to stump like charged electrons. They drop so rapidly to the ground, it's as if they are sucked there through a vacuum. Their descent to the safety of the tangled brash is so sudden it is not really a descent at all, more like a vanishing act.

I cross the valley floor and start to climb through the trees towards the ridge. Buzzards are up hunting the valley. When I reach the top of the ridge I turn west and follow it till I reach the felled clearing. Acres of forest open up beneath me. A crow is calling from its perch on a thin bark-stripped pine trunk. I see it watching me as I enter the clearing and hide myself amongst the brash and stumps.

The next few hours are a dizzying experience. After days spent looking for goshawks and finding only their absence, the birds are suddenly here and all around me. First, there are the usual outriders, buzzard and raven. Two ravens, displaying, tumbling, corkscrewing above the ridge. Three buzzards, calling – also displaying – above my head, twisting, diving, rushing down towards the forest. Then, lower than the buzzards, 30 feet to my right, flying just above the tops of the clearing's remnant trees, a goshawk's blue-grey back. The hawk swings over and hangs huge above me so that I nearly fall over backwards straining my neck to watch it. I can see the contour patterns on its chest, the darker band of feathers round its eye, the white stripe along its eyebrow.

The next one is inside the trees, moving in and out of view. A large shadow flying through the trees, as if they were parting to let it through. Suddenly swinging up to land on a branch. I cannot see all of the bird because it is perched behind a curtain of leaves. There is just a hint of grey shoulder and folded wing, an unevenness of colour inside the trees.

A goshawk spends most of its time perched, blended into a tree. Absence or agitation can sometimes betray the presence of a goshawk: an absence of corvid nests (or successful nests) suggests a goshawk is in residence in a wood, as if the hawk has pulled in an exclusion zone around it. But this absence can just as quickly flip into mobbing agitation, crows furious with fear and ancestral spite, yelling to the wood, to the wind, to the whole world: *Here is a hawk.*

Most birds of prey wear the mobbing with indifference until it becomes so persistent that they are forced to move on. There is an account of one juvenile goshawk being killed by a flock of 200 hooded crows. But mobbing a larger predator is full of risk and there are many more accounts of the harassed raptor suddenly turning and seizing their pursuer. One goshawk I read about was seen joining a raven in mobbing a sea eagle only for the eagle to turn and stoop on the hawk, catching it in its talons. Occasionally, rarely, the hawk becomes the mobsman: Eurasian eagle owls (one of the few species to predate goshawks) have this effect on most raptors. These apex predators are not tolerated by other birds of prey, to the extent that eagle owls are sometimes used as decoys to lure in birds such as goshawks and ospreys for ringing purposes.

Female goshawks are faster in flight, the males have more agility and acceleration. The females are all low-geared power, gulping up the space. Males are slalomers, threading through the trees, twisting after prey.

Often a kill starts from a standstill, an ambush along a quiet forest ride, dropping from a perch, accelerating into a passing pigeon or a noisy undulating woodpecker. Sometimes a goshawk will stalk its prey, tree-creep, pausing to listen and check its bearing on prey that the hawk may be able to hear but not yet see. Woodland raptors (hawks and owls for instance) hunt with their ears as well as their eyes. Folds in the land, hedges, walls: all are used as cover to take prey by surprise. Not the leisurely contour-hugging of a harrier's flight, more a dash along a ditch, everything done at hawk-speed. A flock of pigeons feeding in an open field is approached from a low glide, skimming the fields. Then, when the flock breaks and scatters, the goshawk is beating, bursting out of the glide, rushing at the stragglers.

Above all pigeons. Also: corvids, grouse (especially the displaying males of woodland grouse), squirrels, rabbits, pheasants, though lifting a bird as heavy as a pheasant is a struggle. Occasionally other raptors are predated by goshawks: kestrels, sparrowhawks, honey buzzard broods ... There are records of everything from domestic cats to birds as large as common buzzards. Even male goshawks must tread warily around the larger female gos, which, recent studies suggest, is becoming larger, the size dimorphism between the male and female goshawk widening in certain localities across the birds' range. Where the goshawk's traditional prey, woodland grouse, have grown scarce, the hawks have shifted to different prey species, evolving in size to adapt to this change. The male goshawks becoming smaller to pursue swifter more agile prey; the females, in turn, switching from woodland grouse to hares and growing larger, evolving to better tackle the hares with all their strength and speed and risk.

A hare must rarely pose a risk. But there is the unfortunate goshawk I read about who grasped a hare in one foot and tried to halt the hare's momentum by seizing a clump of vegetation with its other foot. The hare cried out with its witchy screech but its powerful legs kept moving, charging onwards. The hawk instinctively tightened its grip on both hare and plant and the bird, apparently, was torn apart by the force of the hare's momentum ...

Above all, too, a bird of woodland and woodland edge zones. But the goshawk does not need so much deep forest as we tend to assume. Spinneys, scattered copses, neglected churchyards will do fine. Even city parks, playgrounds, just so long as there is a tree in which to build its heavy nest. Some central European cities now have established populations of goshawks (Berlin has around ninety pairs), where the hawks can be seen hunting magpies, blackbirds and feral pigeons through the parks and streets.

Even – of all the treeless places – Orkney, where MacGillivray had reports from his contemporaries of goshawks *not unfrequently* being seen. And, amongst the many bones of sea eagles that spilled from the tomb on South Ronaldsay, as well as the remains of gulls and corvids they also found the bones of a goshawk. What was a goshawk doing way up there on Orkney miles from any wood? Why would a goshawk, along with all those other birds, be interred in the burial chamber at Isbister? Some of the birds – the gulls and corvids – are carrion feeders like the sea eagle and may have also played their part in excarnating the dead. Goshawks will sometimes eat carrion, though are not renowned for it. But all the birds that were discovered in the tomb – the sea eagles, short-eared owls, black-backed gulls, raven, goshawk – have in common that they are long-winged, powerful fliers. So perhaps they were included in the tomb because of their strength and flying prowess. For the dead need things that will be of use to them, coins to pay the ferry-man, amulets, birds capable – strong enough – to assist the soul on its journey.

MacGillivray rises about half past eight and leaves Newton Stewart at nine feeling stiff and slow. A mile from town he hears somebody coming up behind him on the road:

– Are you gaun far this way?
– Yes.
– How far are you gaun?
– East a good way.
– Eh?
– East apiece.

The stranger walks on, hesitatingly, then half turns about, tries again:

– Because if you were for company I'm gaun that way too.
– I walk alone always.
– Eh?
– No.

At Dumfries MacGillivray goes in search of Robert Burns's grave. He wanders about the town in the dusk looking for the churchyard. Struggling to find it, walking round in circles, he almost gives up and heads back to the inn. But then the thought strikes him, he cannot possibly pass up the opportunity to pay his respects to the poet. So he wanders on and at last sights the kirkyard. The gate is locked so he scrambles over the wall and drops down into *a wilderness of tombs.* Burns's tomb is locked with a sign from the magistrate warning that he will throw anyone in jail who is caught climbing the wall. *What*, thinks MacGillivray, *have I to do with restraint? These walls were not intended to exclude me, for the memory of the poet is dear to my heart, and I could not injure his monument.* So he climbs the railings, sits down on the steps of Burns's tomb, and bursts into tears.

It is nearly dark now and MacGillivray spends the next hour on the steps of the tomb in a muddle of grief. He feels that the poet's memory is dearer to him than any living being. He doesn't know why this is, only that there is a kinship, a recognition there. It is not that Burns is also a child of nature, rather that MacGillivray feels acutely Burns's own misfortunes and his untimely death. He tries speaking to God, who he has not spoken to for half a year. Afterwards he feels a soft melancholy, for often, he writes, *there is a stage in the paroxysm of grief which to me is highly pleasing.* Then it is past: MacGillivray climbs out of the tomb, back over the cemetery wall and drops softly down into the street.

* * *

It was not always easy to follow the line of the Border accurately. This middle stretch of the boundary is so detached and abandoned, swallowed by forestry. Left to its own devices the line behaves strangely here. It jinks and doubles back on itself, rushes off at sudden, unexpected angles across the hill with no thought to topography. I trace the Border line between Deadwater and Hobbs' Flow on my map. Deadwater, Foulmire Heights, Bloody Bush … the place-names do their best to ward you off. Even Hobbs' Flow comes with a warning from the great explorer and chronicler of the Border, James Logan Mack:

> In a wet season the passage of the Flow should not be attempted, and even in a dry one the traveller is not free from the risk of being engulfed in the morass. While I have crossed it twice in safety, I do not advise that this route be followed, and he who ventures into such solitude should keep to the west and circle round on higher ground.

If you went on the run from justice in Scotland or in England, this is where you ran to, the Debatable Grounds, the Batable Lands, the No-Man's-Land that straddles sections of the Border. A place which neither nation could agree on, a refuge, a place to flee to. *Batable*: contentious, discorded lands. To *bate*: the austringer's term for an untrained, skittish goshawk when it tries to flee the wrist. *Batable*: the reason this middle section of the Border above Kielder behaves so strangely, jinking, doubling back on itself, given to sudden unexpected flights. Maps from the eighteenth and nineteenth centuries describe much of this area as 'Disputed Ground', uncertain, feral places, tending to bate, to flee the authorities. The courses of streams that marked this section of the Border were often diverted, dykes ploughed up in the night, lines erased, land appropriated. It takes a surveyor to referee such disputes, to compromise, to mark a boundary so esoterically with little or no thought to topography. Only a surveyor could create such sharp angles, draw such straight and unexpected lines across the hills.

I climb along the flanks below Deadwater Moor. *Deadwater*: still water, water with no movement; a watershed, between east and west, between the North Sea and the Solway. Also, a watershed for language, above all dialect. Just as in the Flow Country where Norse and Gaelic met and barely recognised each other, the Anglo-Scots border cuts language markedly in two. The distinction between dialects on either side of the Border is abrupt, like nowhere else. MacGillivray felt this when he stopped in Carlisle to ask the way and found the dialect so completely altered:

– When thou comes to the corner, though maun keep to the right, and when thou comes to some houses they will show thee.
– I thank you.
– Welcome.

I treated the Border as the most porous of things, to be breached constantly on my walks along it. But language – dialect – slams up against it, reflects the presence of a boundary along almost all of it. The whole district is rife with dialect isoglosses. Within the space of a mile or two you can go from being a *scarecrow* to a *flaycrow* to a *crowbogle* to a *tattiebogle* ...

The Batable Lands had their own rules. From sunrise to sundown they were treated as common ground where livestock from either nation could be grazed. Anything left overnight – cattle or goods – was fair game and could be taken or destroyed. If property was built inside the area it could be burnt down and any people found inside the building taken prisoner. From time to time the district would be settled and in turn provoke the authorities to clear the people out again. This was the objective behind the Treaty of Norham of 1551:

> Debatable Ground should be restored to its old condition, according to usage, that therefore all subjects of either realm living in it or having houses there were to remove with wives, children, goods and cattle by the Feast of St. Michael the Archangel next, any found there after that date to be expelled by the Wardens and punished according to law.

But always there was a drift of people returning to fill the vacuum, fugitives, rebels, the dispossessed. As quickly as the Debatable Lands were cleared they were colonised again.

* * *

Goshawks went the way – the same ways (and at the same time) more or less – the osprey and sea eagle went. Bounty schemes, vermin lists, clearances ... When they began to return in the 1960s goshawks did so more quietly, more secretively than the sea eagle and osprey: falconers releasing – sometimes losing – goshawks to the wild. A slow trickle of imports and escapees, the latter often trailing jesses into their new lives. The birds are well established now in certain districts, especially the Anglo-Welsh and Anglo-Scots borderlands, but there remain areas, debatable grounds, where goshawks are expelled before they have arrived, sink holes, briefly colonised before they are cleared out again.

My last hour in the forest. The buzzards have returned, the raven too. Late afternoon, light going around the edges of the clearing. The raven steadies out above the treeline and starts to rise and fall in peaks and troughs. Then a goshawk is calling and lifting up from the dark trees to the left of me. This hawk has been calling on and off for much of the day. I am glad to see her. She is much larger than the male goshawk that flew over earlier. Wings held straight and stiff, she flies around the rim of the clearing then crosses in front of me. Slow wing strokes: glide, flap.

VII

Kestrel

Bolton

Late October on the outskirts of Bolton, everything drenched in mist. The town below me in a sump of fog. Rush-hour buses sliding down the hill, tail lights like fading fire ships in a haar. I cross the main road and skirt around the back of a pub. Dormant extractor fans, a yellow drum of cooking oil, starlings in the beer-garden grass, very black in the pale light. Behind the wheelie bins, a five-bar steel gate, an electric strap fence hooked to the gate's straining post. The white strap runs out across the field, 3 feet off the ground, undulating, held up by yew-green plastic pegs. I climb the gate and the metal rattling sound it makes spooks the starlings into flight, black shots through the mist as if the birds were punching holes in it. Beyond the gate a field of deeper mist, the ground bumpy with horse dung, a grey fur on the dung like frost. The strap fence branches off at right angles from itself, demarcating the field into tiny paddocks. The field is full of docks, small thin rabbits bolt from under their blotched leaves.

Another crossing point, another border post. The switch is so sudden here, from town to moor: the end of a street, a sliver of no-man's-land, gorse and bramble, a wet field overrun with rushes. Then I am climbing up onto the crumbling, hacked-at moors. Farms perched on the edge of huge quarries, deep bowls of mist. A peregrine down there somewhere, I can hear it calling from the quarry workings.

Above the quarry, a large field sculpted by landfill, buried and smoothed over, its surface pitted with rubble, glass, tight balls of burnt plastic. Damp sheep on the landfill, muddy Herdwicks, hooves clacking against bits of brick. Across a ditch and the next field is soaking wet. A stunted, windswept holly along its edge, remnant clumps of heather rolled up against the windward side of the dyke, rushes in the peaty hollows. Snipe are here, bursting out from under me in rapid, jagged flight. And from out of the mist a wonderful sight: a large flock of lapwings, thirty birds, rising from the soft field with such lightness, swooping over the moor, flickering black and white. I keep on disturbing them in the mist, dislodging the flock, watching it rise and settle again further up the field like a thrown sheet coming to rest over a bed.

Then something large is heading towards me out of the mist. A tall heavy shape and I think it must be a horse made skittish by my presence in the fog. It is trotting towards me and I am worried that it will not see me in time, that it will rear up and bolt in fright. But when it steps out of the mist and is almost on top of me, the shape and movement are wrong. Not a horse at all: nothing like a horse. What on earth … an ostrich! Of all the places, in the middle of this boggy field above Bolton, striding out of the mist, skidding to a halt in front of me. Thick neck, head tilting down, peering at me. Its huge brown eyes blink. Then it wheels around, its feet slap a puddle left by a tractor print, and it strides back up the hill into the mist. Something about the bulbous shape of its body, the loose straggling feathers, reminds me of huge buzzard nests I've seen, like a nest on stilts. So now, I think, I have the avian extremes of my journey: the tiny wren that bobbed around my heather lookout on Orkney, to this escapee ostrich, fugitive giant of the West Pennine Moors.

I left MacGillivray in Carlisle, asking directions for the road south.

– When thou comes to the corner, though maun keep to the right, and when thou comes to some houses they will show thee.
– I thank you.
– Welcome.

Beneath the sandstone glow of Carlisle castle MacGillivray goes into a bank and exchanges his five Scotch notes for English ones. He searches among the back streets for cheap lodgings but there is nothing available and no one will give him change for one of his notes. So he heads out of the city along the Keswick road. It is a shock how sudden the dark comes on, there is so little preamble. MacGillivray arrived in Carlisle at five o'clock with no sense that the day might be dissolving. By seven he is out on the road to Keswick and there is not a trace of daylight left. The night is so assertive now.

He does not say why, but the next inn refuses him lodging. Though his appearance cannot have helped: he looks so tattered and road-weary, like some wandering scarecrow come down out of the north. Half a mile on he comes to another inn and it's, *No beds here, mister.* But he orders supper anyway because by now he is exhausted, has not eaten a thing all day except for a twopenny cake. But here is a new difficulty which he could not have seen coming: the innkeeper can give MacGillivray change for his note but she cannot accept a Bank of England note. And no, it's no good that MacGillivray only exchanged it at the bank in Carlisle a few hours ago. She is very sorry and her husband is sorry too, you do sound like an honest fellow but, you see, there are Bank of England forgeries circulating through Cumberland and, you must understand, they have been cheated once before and resolved to never take a Bank of England note again. It is too much, this small injustice, and MacGillivray hears himself almost yelling that he is hungry and tired and can scarcely proceed upon his journey. And, yes, you do seem like an honest fellow, really, and they are very sorry, but it will not do. So MacGillivray is back out through the door and trudging along the road. And something about the rhythm of walking once more – reverting to what his muscles know – calms him. But it does not alleviate his hunger and he is brittle with tiredness. It is cloudy with slivers of moonlight showing between the clouds. The road has narrowed to a lane, tall hedges loom up on either side. The lane is exceptionally muddy and MacGillivray is soon wet to his ankles. He notices, when the lane sinks beneath the hedgerows, that he is walking through a seam of shadow and he sees how the night's gradient can change so quickly as he passes through pools of darker, cooler air. Crossroads keep bisecting the lane. Soon he is sure he is on the wrong road and has to knock on the door of a house to ask directions. Further on he comes to another inn. But it's no beds here, mister, and, sorry, but we can't let you sleep in the outhouse either and the cook has gone to bed and so it's out the door again, first turning on the left, then right, then straight ahead to the common. Another crossroads, another farmhouse. The door is shut. MacGillivray finds himself swaying in the centre of a quiet farmyard, a cart shed in front of him. So he goes into the shed, climbs up onto the cart and lies down. Then he notices steps up to a loft, so he climbs up and there are rushes and wattlings plus a bit of scruffy mat he has carried up from the cart. Unhooks his knapsack, punches it into a pillow, lies down on the rushes, drapes the mat over his legs, tries to sleep. At midnight his wet feet wake him with cold. Puts off his shoes, wraps his feet in the mat. Then sleep of sorts till he is woken by people moving below in the farmyard, feeding the horses and poultry. Descending the hayloft, back down into the muddy lane. He seems to have passed beyond hunger into a dizzy weakness. He is, by now, covered in grime as if a carriage had splattered him with mud as it hurtled past on the road. And he must have slept on his hat because it is so crumpled it will not be coaxed back into shape. Two miles down the road and he enters a public house. Breakfast at last: tea, bread, butter and eggs. And MacGillivray does something which is unlike him and which racks him with guilt for days to come. He waits until he's finished his breakfast before he presents, nonchalantly, his Bank of England note. Both the innkeeper and his wife, inspecting it, prodding it, shaking their heads. Sorry, but if you had a Scotch note instead … they would be happy to give change for that. But a Bank of England note: no, we simply cannot take the risk, would rather forfeit the shilling for your breakfast. And they show MacGillivray great kindness, this couple, and say that it does not matter and he is welcome to the bread and eggs. Still, he leaves them feeling a little guilty, a little sly he had not mentioned the pesky notes up front. He makes a vow not to be so dishonest again. Outside the inn he can see Carlisle cathedral in the distance. The hills of Scotland are rinsed in mist. It is eleven o'clock in the morning. The mud on his clothes is starting to dry and flake.

> **Kestrel**
> There it comes, advancing briskly against the breeze, at the height of about thirty feet, its wings in rapid motion, its head drawn close between its shoulders, its tail slightly spread in a horizontal direction, and its feet concealed among the plumage. Now it sails or glides a few yards, as if on motionless wings, curves upwards some feet, and stops short, supporting itself by rapid movements of its pinions, and expanding its tail. In a few seconds it flies forwards, flapping its wings, shoots off to a side, and sails, then rises a little, and fixes itself in the air. On such occasions it is searching the ground beneath for mice and small birds, feeding or reposing among the grass. Having discovered nothing, it proceeds a short way, and again hovers. In a few seconds it wheels round, flies right down the wind at a rapid rate, to the distance of some hundred yards, brings up, and hovers. Still nothing results, and again it glides away, bearing up at intervals, fixing itself for some seconds in the air, and then shooting along. When about to hover, it rises a few feet in a gentle curve, faces the wind,

> spreads its tail, moves its wings rapidly, and thus balancing itself keenly surveys the ground beneath. The range of the tips of the wings at this time is apparently about six or eight inches, but sometimes for a few seconds these organs seem almost, if not entirely, motionless. The bird has once more suddenly drawn up, and is examining the grass with more determinate attention. It slowly descends, fixes itself for a moment, inclines a little to one side, hovers so long that you may advance much nearer, but at length closing its wings and tail, falls like a stone, suddenly expands its wings and tail just as it touches the ground, clutches its prey, and ascending obliquely flies off with a rapid and direct flight.

How does a kestrel hover the way it does? It is a movement of such precision and control. The falcon holds itself in the air like a star you could navigate by. The bird is standing still, walking into the wind. Its head, its eyes are fixed to the ground beneath. Every minute adjustment of its wings and tail is made to keep the head in place. It is astonishing it can keep its head that still despite the wind which flings itself at the bird's light frame. And really – more accurately – the kestrel is not hovering but balancing itself, pushing itself against the wind so that the force of the wind is cancelled out by the momentum of the bird's flight. The kestrel is held in place, pinned by the interplay of wind and flight. If it is pushed back a few millimetres by the wind the falcon stretches its neck to keep its head in place. It adjusts to the rise or fall in the wind's speed by alternating between glide and flap.

To hunt like this the kestrel needs the wind to hold it up. But if the wind is too strong, it becomes difficult to hover effectively, even more so when there is no wind at all. The kestrel can hover on a windless day, wings flickering, winnowing the air. But it is expensive – inefficient – to keep this up. So when the wind is not right or, in winter, when energy is precious, a kestrel hunts by other means than hovering. Sometimes it resorts to piracy: dive-bombing short-eared owls to make them spill their catch. But more often the kestrel will revert to hunting from a static perch, a hedgerow, tree or telegraph wire. A beetle can be spotted clambering through the grass from a perch 50 metres away, starlings foraging amongst cattle are clocked at 300 metres. It is less successful, this method of hunting, because the falcon cannot cover as much ground. But, especially in winter, static hunting conserves crucial energy. In spring and summer, when the male kestrel needs to work overtime, more often than not he reverts to aerial hunting.

After the ostrich and the field of mist I climb higher up onto the moor, but it is hopeless up here as I can see very little in the fog. A kestrel could not hunt in this. Occasionally the mist shifts enough for me to see a flashing red light from a television signal mast on the summit of the hill. The narrow road leading to the mast is cracked and split where frost has picked at it. I head down off the hill and follow a path pebbled black with sheep droppings. A few sheep cross the path ahead of me, smudged identification marks sprayed onto their backs, the turquoise dye corposant on their wool. On the lower slopes there is a brightness building inside the fog. Then suddenly I am out: the mist has ended like a stratum of the moor and I can see the wooded slopes above Rivington and a great stretch of the Lancashire Plain opening up below.

Of all the birds of prey the kestrel flares the most with light. In Gaelic she is the *Clamhan ruadh*, the red hawk. *Ruadh*: a less true red than the bolder, purer red of *dearg*. *Ruadh*: the coppery red of fox and rust and a roe deer's summer pelt. The male kestrel is the brighter red, his brightness offset by his beautiful pearl-grey head and tail. The female lacks this grey but both have black wings from the carpal joint down to the tip which accentuates the ochre colour of their back and upper wings. The female's red is thinner, browner, more chestnut-coloured. Often, when the falcon banks to glide away, you catch that blush of colour down their back and wings. Sometimes, when the sun brushes a kestrel's back, the bird's feathers glint with brightness.

One night in July 1913 a huge blaze is spotted on the side of the hill above Rivington. Lord Leverhulme's wooden summer home is on fire, burning through the night. A few hours earlier, the suffragette Edith Rigby had walked around Leverhulme's property checking nobody was inside. When she was sure the house was empty she broke a window and poured paraffin through the hole, felt the trapped heat of the day spill out of the house when she smashed the glass. The place was tinder-dry, the pitch-pine walls warm to touch. Later, a night breeze licked up the fire, and from across the valley flames could be seen pouring out of the hillside. At her trial Rigby said:

> I want to ask Sir William Lever whether he thinks his property on Rivington Pike is more valuable as one of his superfluous homes occasionally to be opened to people, or as a beacon lighted for King and Country, to see that here are some intolerable grievances for women.

I walk on down the hill's shoulder. Around the 1,000-feet contour point, on the western edge of the hill, there is a large levelled area cut into the side of the moor. This is the spot where Leverhulme's house stood. The area is matted with reeds and thick tussocks of grass. There are hints of concrete, like puddles, in amongst the grass. The flatness, the concrete undercoat, make the place feel like a neglected car park. No trace of the house, as if it had been swept clean away. Except, I find some black and white chequered tiles amongst the grass, the remains of a lavatory floor (once situated off the entrance to the ballroom). The tiles are in good condition, the white ones even look clean. A few dead leaves have settled on them as if someone had left a door open.

After the fire the house was rebuilt with stone. Leverhulme, who had made his fortune from packaging soap, was one of the wealthiest industrialists on the planet and he poured money into his hilltop home. Photographs from its brief heyday are otherworldly: wooden pagodas beside a Japanese lake, a glass-roofed pergola, a minstrel's gallery above the ballroom, fluted pillars, Flemish tapestries, an 'orchestra' lawn. An entire hillside reclaimed from the moor and sculpted into ornate Italian gardens, boating lakes, waterfalls, terraces of rhododendrons. When Edith Rigby was reconnoitring the house for her arson attack she walked through the grounds passing grazing herds of llamas, kangaroos and ostriches. She noted how the animals seemed better housed and fed than many people on the land.

I stand up from the sundial stump where I ate my lunch and start to drop down off the hill, walking through the gardens. Everywhere the moor has crept back in. The croquet lawns are thick with rushes, the lakes silted up and grown over, the summer houses in brambles. Along the terrace steps, in the Italianate archways, you can see the detailed craftsmanship in the stonework. Then I am out of the terraced gardens onto a change in gradient. The ground begins to level and I think this is where the zebras and ostriches must have grazed, where Edith Rigby parked her car before the long slog up the hill with the paraffin kegs. The paths here are deeply grooved by runoff from the moor. Allotments, school playing fields, the woods giving way to the creased edge of the town.

* * *

A few years after Edith Rigby burnt down his home, Leverhulme bought the islands of Lewis and Harris in their entirety. The village where William MacGillivray went to school as a boy, Obbe, Leverhulme renamed after himself, changing its name to Leverburgh. A century on from the meeting between MacGillivray, his uncle and the laird over the tenancy of the farm at Northton, the same land disputes continued to flare through the islands. In 1919 a group of crofters from Northton challenged their new landlord, Leverhulme, with his lack of empathy for their need for land, with the same passion as MacGillivray had done when he confronted MacLeod and his duplicitous factor, Stewart. In a letter from 1919 petitioning the Board of Agriculture for the farm at Rodel, ten crofters from Northton wrote:

> We shall never submit tamely like our forefathers. We shall not be compelled to leave our native land without struggle. If something is not done soon, I am afraid we shall be compelled to take possession in our own way. The following are those who wish to be given a small-holding on Rodel without delay …

When MacGillivray opened up the kestrel to inspect its stomach he found, for the most part, the hair, bones and teeth of mice and shrews. But also, he wrote,

> I have found the remains of young larks, thrushes, lapwings and several small birds both granivorous and slender-billed, together with the common dung-beetle, many other coleopteran, and the earthworm.

Add to this: the occasional lizard, rat, mole and slug (one kestrel was observed carefully skinning slugs, lifting them up to its bill, swallowing the slugs whole, a residue of slime glistening on the kestrel's foot). The kestrel is somewhere between a specialist and generalist, somewhere between the short-eared owl, with its dependency on voles, and the more omnivorous buzzard. The kestrel can adapt its diet, diverge from voles when it needs to, but field voles are its mainstay; it cannot completely do without them. Population density, breeding success: both are impacted by the availability of voles. Starvation, the most common cause of mortality in kestrels (juvenile birds especially), is most prevalent in poor vole years, or when the voles are sealed by heavy snow.

They live short lives, these small falcons. There is none of the slow burn to adulthood you find in larger birds of prey. Kestrels can breed in their first year. The majority die before their second. They can be sedentary or migratory; much, again, depends on voles. There is an emptying out of kestrels from the uplands in winter; some birds migrate long distances, others remain in their breeding territories year round. Nest is a hole in a hollow tree, a disused crow's nest, a scrape on the side of a cliff (like all falcons, the kestrel does not build a nest). On treeless Orkney, where there are no foxes, kestrels nest in deep heather. Buildings serve as stand-in cliffs; in German the kestrel is the *Turmfalke* (the tower falcon). Window boxes, windmill ledges, church steeples, gutters ... as long as the site is sheltered, out of reach of ground predators, and will contain and hold their eggs.

There is a risk we take this common bird of prey for granted, that we view the kestrel as more adaptable than it is capable of being. The kestrel is not as common as it used to be. Since the mid-1970s its population has declined in England by more than a quarter. Scotland's population has also plummeted in the last two decades. If we destroy the habitats for field voles, the meadows and field margins, the tangled unkempt places, kestrels will find it hard to cope.

From the town I keep looking back towards the moor and the wooded slopes below Leverhulme's home. It is not so clear-cut, the switch from moor to town: the two can mingle and fuse with one another. Kestrels move easily between the two spheres, hunting over the moor, nesting on the town's buildings, on its disused factory chimneys. And sometimes the town slips out of itself to re-emerge up on the moor. In September 1896, 10,000 people walked up through the streets of Bolton, climbing up onto the hill in a mass trespass to stake their right to follow a path over the moor, the buildings emptying that morning and the noise of the exodus like a slow exhalation of the town's breath. Gamekeepers and policemen met the trespassers at the edge of the moor, noting ringleaders, the crowd churning past them through the disputed gate.

During the war a replica town was built up on the moor: a decoy, constructed of lights and long channels laid over the turf. The lights were used to resemble a poorly blacked-out town, leaking light from its doorways and factory furnaces. The purpose was to fool enemy aircraft into thinking they had reached their target. Fuel in the channels was ignited then flooded with oil and water so that it looked as if a town was already burning (and exploding) down there. The decoy was built on the moor to protect the crucial royal ordnance factory near Chorley. The fuel tanks, pipes and poles with lights on top of them stayed in place up there, on the hillside to the west of Belmont, for the duration of the war. Not so much a replica town, more the idea – the thought – of a town sketched onto the moor. A town stripped down to its circuit board.

I am walking through the suburbs of Horwich now, heading towards the M61 motorway. The hum of its traffic has been with me all morning. On one street, in a tiny front garden, there is a birch tree with all its leaves shaken out. A collared dove is perched in the tree like a bulb of grey light. MacGillivray paused near here on route to London, recharging his spirits in a public house in Chorley, sitting beside the fire, smoking a pipe, deciding then to not give up, to shun the stagecoach seat, and carry on with his walk to London, *although*, he wrote, *my stumps should be worn off to the knees ...*

I turn left off the street into a thin alleyway that runs along a row of back gardens. Old grass cuttings, tipped over the fence, are suspended in bramble bushes like frozen spray. The ground is slippery from the mulch made by the cuttings. In places logs have been laid across the path to serve as stepping stones. The other side of the path is flanked by the embankment of a disused railway, its slope rusty with discarded Christmas trees. Birch and hawthorn take over as the path becomes muddier and I drop down towards the edge of the town.

The rest of the day I spend walking beside the motorway through the wet fields on either side. This is where I hope to find the kestrel, in the tall grass along the motorway's sidings, between the slip roads and roundabouts. I use the footbridges over the motorway as lookout points, scanning the embankment with my binoculars, swaying above the traffic. A farm track leads to a tunnel under the motorway; it is a quiet echo chamber, lit with puddles, with a high ceiling to allow farm vehicles through. The tunnel is a good place to pause a while away from the wind and whoosh of the motorway. Its arched entrance frames the view of where I have walked today: the hill, now clear of mist, the wooded slopes of Leverhulme's old home, the town across the fields to the east.

If you transplant something ingrained in a landscape to somewhere new, can the object still retain the semblance of itself? Or, once it has been removed, does it start to die a little in its new surroundings? When I was on the Moray Firth I did not feel I could gather up the stones I found along the shore because it seemed that to remove them would somehow have diminished them.

There is an anecdote about Lord Leverhulme, that he took a liking to a wooden fireplace mantel in a public house on the Isle of Lewis. It was an ancient slab of oak and for generations families had carved their names in the wood. The owners of the inn would not part with it. So Leverhulme bought the entire building instead, extracted the mantel and shipped it down to his summer home at Rivington, where he had the mantel installed to support a fireplace alcove in the ballroom.

The stone used to build this motorway tunnel was transplanted from the moor above the town. Thirteen feet of stone were skimmed from the hillside to lay the foundations for the motorway. And it's almost as if I have not left the hill at all, that I'm still walking over the moor – or a memory of it – as I criss-cross the motorway. The stone removed from the moor was replaced with soil which washed off the hillside in the next heavy rain, pouring down the lanes and spilling over the main road in a thick black waterfall.

Nothing is static: even a hill can suddenly burst and wash half its face away; a bog is always breathing, changing its shape in its sleep; a kestrel appears to be still as she hangs in the air but really she is moving, pushing against the wind.

This is where I hope to find the kestrel and this instead is where I find the buzzard, flapping low over the scrubby no-man's-land between the railway and motorway. And then, because buzzards will often do this – will often draw your eye to something else – I spot the kestrel, hovering above the junction roundabout. I get as close as I can to the motorway and prop myself under an oak tree overlooking the slip road and roundabout.

The kestrel holds herself at a steep angle against the wind, tail spread to increase the lift, wings shivering. Her strikes come regularly: once, often twice, a minute. A staggered descent through the air, the angle almost vertical, checking her drop 5 feet above the ground, a slight adjustment to left or right, then a final plummet into a splash of grass. Those last seconds, when she pulls her wings up behind her before she hits the ground, I can see the pale white flash of her underwings and tail.

When the kestrel leaves the roundabout she heads south-east along the motorway. I scramble after her and the last few hours of daylight are a rush along hedge-rows, muddy underpasses, splashing across fields. There are snatched glimpses of the kestrel, often in the distance, hanging above the motorway. To catch up with her is just to push her out of reach again. Half a field away is as near as I get before she banks and arcs across to the adjacent field. In a farmyard I am slowed down by a herd of cattle, their breath smoking in the cool air. Curious and nervy, they bunch and heave around me, make sudden splatter-ing retreats through the mud, remuster and hunker back towards me in a swaying huddle, heads low to the ground, huffing, yellow tags hanging from their ears like bright fruit.

As the sun is going I am in a field above the motorway. Jackdaws and rooks are heading to roost in ragged flocks, the smaller jackdaws weaving amongst the rooks, the molecular structure of the flock. I am watching the kestrel perched in a hawthorn bush, her folded wings twitch and jerk above her long tail. The motorway is a streak of yellow light. The low sun has brought out the orangey reds of the kestrel's back. She is surrounded by the darker, deeper reds of hawthorn berries. I stand behind her and watch her shape turn black.

VIII

Montagu's Harrier

The Fens

One night, on route to London, MacGillivray dreams of flight:

> I dreamed the other night that I was winging through the air in a large area about three or four feet from the ground with great velocity, and I felt so very happy that I scarcely remember to have ever felt

> happier ... the impression which this aerial tour made upon my mind was so strong that for some time after I could not prevail upon myself to believe that it did not actually happen and I can scarcely believe that it is not possible.

All flight, dreamed or otherwise, falls short of the bird I am watching from this hedge. Montagu's harrier: a bird given over to buoyancy and lightness of drift. Flying, like MacGillivray dreamed he did, 4 feet from the ground. An adult female, working her way down the purlieu of the hedge, coming towards me in lilting flight. All wing: there is nothing there but wing and nothing for the wings to do but stretch and glide. I have never seen such buoyancy, never known something so undeterred by gravity. That she does not stall, flying so low and slow above the ground, is a miracle of design. And there is no wind today, nothing to hold her up, just heat and a thick stillness clogging the day.

I am on the edge of The Fens. It is the middle of July and the air is furrowed with heat. I have found a place inside the hedge which allows me to swivel easily and look out over the fields on either side. The hedge is tall and frayed, more like a loose phalanx of trees. There are deer gaps and fox gaps and some openings that are so wide they could be sluice gates left open to let the wind rush through unimpeded. The bulk is hawthorn and hawthorn's companion, elder. There is also blackthorn with its darker, glossier leaves. And a dead oak, cloaked in ivy, standing up to its knees inside the hedge. The fields on either side are wheat and barley, the wheat greener, paler-looking, the barley crackling and popping in the dry air.

When MacGillivray came down out of the north on his walk to London, he suffered a sort of vertigo – an inverse vertigo – or rather, a dry form of the bends, descending too quickly from the northern uplands. It left him homesick and irritable. His instinct, like a ptarmigan's, was to cling to the high ground for as long as possible.

In Keswick his Bank of England note is changed at last. He heads out of the town on the Borrowdale road and the rain comes on again. He stops to ask directions for the way to Ambleside but ignores the advice that he should return to Keswick to take the regular road as it goes against his determination to see the mountain. Because mountains will do that to him: pull him to them, not for the sake of climbing, but because he wants to see what is growing on their slopes, what alpine flora can survive up there. Sometimes it feels as if every plant – or the prospect of a plant he has not seen before – has this magnetic pull on him. And I cannot fathom, with every mile along the road an infinity of distractions, how he is able to walk to London as quickly as he does.

So MacGillivray follows this new diversion and heads up on the footpath over the fell. On the outskirts of a village a dog rushes at him with such fury he has to yell at it to get back. The noise of its barking – and the sound of his shouting – is such a shock after so many miles of silence. And the dog's sudden burst of fury is not unlike the London juggler, *driving like fury at his squeaking fiddle*, whom he sat beside in the Keswick inn, falling foul of his rum, trying to dry out in front of the fire after his shivering night in that hayloft outside Carlisle.

Still raining as he enters the valley and he can see the rain on the mountain *glistening on the face of the dark rocks*. He feels himself being ballasted with tiredness. Sometimes he feels so tired he does not recognise himself, as if a part of him has travelled on ahead and knowledge of who he is, or that absent part of him, grows blurred and strained. He struggles, for instance, to remember his age, as if he had mislaid it, or mislaid a year. It plagues him for mile after mile, this dizzy insecurity, and he is only rid of it – this giddiness – when he grasps that what is really happening is that all this time alone on the road has left him, despite what he says, unsure of his own company, of who it is that is walking in him. He finds it hard to recall the person who left his house in Aberdeen a month ago. He would like to ask him, his distant self, what on earth it is he is doing scrambling up this mountain in the near-dark in Cumberland. At such moments he wants to abandon the walk, return to Scotland, cloak himself in study, *become useful* ... But he has walked so many miles over the years, back and forth from Harris to Aberdeen and all the digressions in between. So he knows too well that walking will do that to a person, that something in the friction of momentum, in the pounding of the road, unsettles the soul so that it becomes frayed and loose till it drifts up and away from the body like a kite. And it's at that moment, when the soul hangs flapping above the body, that he loses sight of who he is and finds himself running through the check-list of himself, fumbling, struggling to reconcile himself and haul back in the errant soul.

He reaches the summit of the mountain at dusk. The plants here: savin-leaved club-moss, prickly club-moss, common club-moss, fir club-moss and the starry saxifrage. He has, by now, lost the path completely. So he heads down the mountain through the closing dark, picking his way across the scree.

A whitethroat scolds me when I arrive at the hedge. It keeps leaving and returning and each time sounds a little less agitated. The Montagu's harrier's nest is in the deep barley field in front of me. For much of the day the female harrier has been circling above the nest. Occasionally the sun catches and brightens the dim white colours in her wings. There is no sign of the male. He has another nest – another mate with young – a mile away in a dried reed bed. I would love to see him, the ash-coloured male. MacGillivray wrote that the male Montagu's harrier was *remarkable for its slender form and the great length of its wings*. He is the lightest of all the harriers. He has the largest wings compared with body size of perhaps any bird of prey. Low body weight accentuated by long wing length: this is the ratio which gives the Montagu's harrier its extraordinary buoyancy, its ability to keep on sailing out low over the ground for mile after mile. The birds are all wing and can easily spend half the day on the wing. At a steady 20 mph gliding over the land, the distance covered in a day by the Montagu's harrier could be anywhere between 50 and 100 miles.

He is not a static hunter. He is nothing like the kestrel who waits up inside the wind, or the patient goshawk who sits in ambush on its forest perch. The Montagu's harrier hunts on the move, a low-level, long-distance forager, drifting out to scour the land. Like other harriers, Montagu's hunt by sight and hearing, using the cupped disc shape of their face as an owl does, funnelling any noise they detect onto their ears. The Montagu's harrier's face is a listening device and, flying as low as they do, they can pinpoint a locust by the sound the insect makes feeding on the branch of a cotton bush.

Lightness and lift: if he finds an updraught, the male Montagu's settles into it, gives his light frame over to the rising current of air, so that he is carried up and away like pollen. In this way he is able to reach hunting grounds miles from the nest site. For such a large bird, if you picked him up, his pumice lightness would startle you. I heard a story of a male Montagu's harrier found floating in The Wash one year after he collided with a wire. When the drowned bird was pulled out of the sea there was nothing to him: he was all feather.

I am willing the male Montagu's harrier into view. It is easy, when you are straining for a glimpse, to over-anticipate the bird, to try to make its outline fit onto a different bird. I have often tried to squeeze goshawks into buzzards, sparrowhawks, even, into fast-flying pigeons. Late in the morning a marsh harrier floated inland, a male, coming in across the barley field, and I missed a beat thinking it might be the male Montagu's harrier returning to the nest, till the marsh harrier's harlequin colours took hold, creams and blacks and reds. The marsh harrier flew close to the Montagu's nest and I skipped another beat because the female Montagu's was away from her nest and, in her absence, I wondered if the marsh harrier would swoop down and try to take her young. But the marsh harrier flapped on, following the line of a hedge, black wing tips against the yellow field.

After the marsh harrier had gone, the female Montagu's drifted back across the field. She stood in the air above the nest for a while, circling there. Then she turned, drifting quickly away. A glider, light as balsa, floating across the fields. The narrow line of her outstretched wings met and matched the line of the horizon, and as she moved further away from me her outline began to shimmer in the haze. There was a gold tint to the feathers on her underside. Then she disappeared in the heat and when I picked her up again she was beside a wood two fields away, her shape brought back into focus by the dark trees. She was climbing, and as she cleared the top of the wood a buzzard rose from under her, heavy and broad-winged and flapping to heave itself up. The harrier began to rise in a wide corkscrew pattern above the trees. The sky over the wood was very bright. As she circled there, the white bar of feathers joining her tail to her body seemed, against the backdrop of white sky, like a gap opening up across her middle. For a few seconds, her tail looked like it was chasing – trying to rejoin – her body.

* * *

Montagu's: after the English ornithologist, George Montagu (1753–1815), who was the first to accurately describe and identify the harrier as a different species from the hen harrier. His identification came after centuries of muddle and it can still be difficult to tell the two apart. The adult male Montagu's harrier is a slightly darker, slightly dirtier grey. A black bar runs across the middle of his upper wings, one bar on each wing like a reflection of itself. Two parallel grey-black bars run along the underwings. His chest and the underside of his upper wings are flecked with a rusty patterning; the female Montagu's also has this reddish-brown blotching on her feathers. The ruff (the wreath of feathers that circles the head) is less pronounced in both sexes of the Montagu's than it is in the hen harrier. The male Montagu's is a more intricately patterned bird, his underside daubed with reds and browns and blacks; the dots of red look like a street of distant lights. The male hen harrier, by comparison, has a clean white undercarriage, unpatterned, brightly lit.

All of these variants are mostly undetectable, even through binoculars. It is easier to look to the bird's shape and the feel of its flight. Look especially at the shape of the wings: the Montagu's has longer, narrower wings with three long-fingered primaries; the hen harrier's wings are broader, blunter at the ends. The difference in wing length is especially noticeable when the birds are perched: the Montagu's long wings extend as far as the tip of its tail. Gravity, as well, works differently on the two species. The hen harrier, with its longer legs, stands taller. The Montagu's harrier, at rest, has a lower centre of gravity, sits more squat to the ground; in flight, gravity seems to have no bearing at all on the Montagu's harrier.

In the early afternoon high cloud begins to collect, thin streaks finding then plaiting each other. There is a slight shift in the light as the day's glare dims under the cloud. When the female Montagu's harrier rises high above the field the backdrop of cloud articulates her outline beautifully. Her three long primaries are clearly visible, the first time I have been able to count them accurately. When she holds her wings out behind her, before she goes into a glide, the tips of the wings have a falcon's sharpness.

I spend the whole day not getting used to her. Each time she comes into view I am amazed by the length of her wings. She is so loose and willowy, so relaxed in her flight. And unlike any other raptor I have seen in the way she does not flex her speed and strength. She is more like a gull in this respect, languid, unhurried in her flight. She keeps her speed in check: an adverse wind specialist, flying into the wind to slow herself down so she can scan the ground. Until the moment she spots her prey and you see how in fact she is primed with speed, is capable of a sudden shock of speed and will often carry on past her prey then turn rapidly around, backtrack, and use the wind's momentum to make her strike.

Striking at: skylark, meadow pipit, lizard, partridge chick, reed bunting, common vole, field vole, grasshopper, cricket ... In their African and Asian wintering grounds, locusts are an especially important prey species for these harriers. The Montagu's harrier can switch between small passerines and small mammals and, though it is not as dependent on voles as some other birds of prey, there are locations where voles are a significant part of the Montagu's diet and the harrier's breeding success is strongly influenced by the vole population.

Lightness and lift, will-o'-the-wisp, the soul set adrift like a plume of smoke ... The poet John Clare described the Montagu's harriers he saw from his home on the edge of The Fens as *swimming close to the green corn*. It is in her lightness and ease of buoyancy that the corn-swimmer is most distinct from the hen harrier, the heather-wanderer, a sense that you have met, in the Montagu's harrier, the epitome of lightness and drift, that you could not perceive any creature more buoyant than this.

If I were to mark a halfway point on my journey I think it would be here on the edge of The Fens. The place feels like a junction of sorts, a crossing point between the north and south. At least, I have the sense that I have crossed a border between birds, between the hen harrier who breeds on the northern uplands and the Montagu's harrier who, on the edge of its northern range in England, breeds in the fenlands and chalklands of the south country. It is a border which dissolves outside the breeding season when, at the end of summer, Montagu's harriers depart for the locust-rich habitats of sub-Saharan Africa, returning again in the spring. In the interim, a change of shift: hen harriers come down from the moors to over-winter in the lowlands, sometimes moving into the vacuum left by the departed Montagu's harriers.

There used to be more interaction between the harrier species in this country, more blurring of their breeding range. Historically the Montagu's harrier has never been a numerous summer migrant to the British Isles, but there are accounts of it breeding alongside hen harriers and marsh harriers in The Fens. There are records of the birds being snared and their eggs sold when the collecting rage was at its height in the nineteenth century (a male Montagu's harrier in good condition could fetch 20 shillings; eggs could go for a similar sum). So this border between birds – this north–south divide between harrier species – is not a natural one. The hen harrier's breeding range is only confined to the northern uplands because persecution forced it to retreat to the far north and west, and it has never recolonised the extent of its former range. In tandem, the Montagu's harrier's breeding range (which historically had been more widely, though still thinly, scattered) has contracted to the opposite corner of these islands and today the Montagu's harrier only breeds in tiny numbers in the south and east of England.

Dissolve the border: bring the different harrier species together, let their territories overlap (as they do across their global breeding range). It is then that you really see how the birds differ. Within a shared habitat each species of harrier is its own specialist. The marsh harrier, with its longer legs, hunts over the taller reed beds; the

Montagu's, with its shorter tarsi and claws, ranges inland across shallower vegetation, pursuing smaller, more agile prey and with a tendency to be more insectivorous than its larger neighbours, the hen and marsh harriers. Even their breeding seasons are staggered so that the different harrier species avoid competing with each other when the demand for prey to provide for their young is at its height. Marsh harrier, Montagu's harrier, hen harrier, pallid harrier (which overwinters with the Montagu's in India and Africa and has a shorter wing and faster flight), they have all evolved to coexist.

Dissolving borders, the interchange of land and water, the blurring of the two together: welcome to The Fens, where the land is lower than the sea and everywhere is peat and silt and water. And if it is not water then it is land reclaimed from water, and land taken from water is land that has a memory of water, or land that tries to dream the water back to life. A place where water loses its identity, where there is no gravity to instruct the rivers, where salt and fresh water meet and meld and sometimes meet so destructively, like two avalanches rushing at each other, that the land is drowned for weeks on end. A place where a plough working a black field miles inland can exhume the skeleton of a whale from out of the peat; where once, during a winter flood, a ship ran aground on a submerged house and, thinking they had struck a rock, the sailors committed themselves to God and saved themselves only by clinging to the roof.

* * *

She is still coming towards me down the side of the hedge. Through my binoculars I can see the cupped disc of her face and the dark bands across her tail. When she veers close to the hedge the white rump at the top of her tail looks like a splash of hawthorn blossom against the hedge. It is the most conspicuous part of her, this white strap across her middle, the most brightly lit part of her plumage, flashing, signalling across the fields. She is moving quickly and I wonder if the lack of wind means she is finding it hard to steady her pace. I try to stay hidden. I don't want to startle her with my presence and so I burrow a little deeper into the hedge, combing the long grass over my back and around my head. My worry is that she will notice the sun glinting off the glass lenses of my binoculars. I watch her as she keeps to the same course, the same linear route across the field.

Lines are an important feature of the way a Montagu's harrier hunts. The birds will often head out from the nest site in a straight line following a hedge or the side of a wood, foraging along its tangled margin, looking to come across prey by surprise. The birds rarely deviate from this flight path, continuing to head out on the same trajectory, sometimes for up to 12 kilometres from the nest site, a much greater distance than hen and marsh harriers forage from the nest. In this way the Montagu's harrier reads the landscape differently from the hen harrier. The latter is more inclined to hunt by hugging contours, dipping and rising with the swell of the land. The Montagu's harrier is a bird of grids and patchworks, of roadside verges and ditches, utilising the linear structures we have drawn across the land.

And there is no landscape more linear than The Fens, no place I have come across in these islands that has been so straightened by drainage and agriculture. Here is an uncovered landscape, the inverse of what happened at Culbin, where the land was smothered by the sandstorm, and the valley of the Upper Tyne, when it was flooded by the building of the Kielder dam.

Before it was uncovered – before it was fully drained – much of The Fens was known as 'half-land', ground that was neither permanently flooded nor high enough to escape the winter floods, a place where the border between land and water was constantly breached and blurred and the boundary between fresh and salt water was often hazy. Sometimes spring tides would breach so far inland that deposits of silt would be left like a smear of grease across the peat. Beneath the peat of the southern fens and the silt zones to the north which border The Wash, there is such an interchange of marine and freshwater deposits layering the soil, it is impossible to tell where the sea began or the rivers ended.

How do you begin to bring such a disobedient landscape under control? It took centuries of incremental progress, of drainage schemes that could be undone by a single night of heavy rain, a drain left unmaintained, or somebody's pigs rooting over a dyke. For every acre of reclaimed land, the fen would revert to water at the slightest opportunity. Often drainage works would be sabotaged by locals fed up with the taxes levied on them to pay for the schemes and the disruptions that drainage caused to their livelihoods, to navigation, fishing and wildfowling.

But the land's potential to be made profitable drove the drainage schemes on. Rivers were straightened and diverted, cuttings and outlets excavated, outfalls dredged of silt. Sluices, locks, scoop wheels, pumping stations were implemented; windmills, then steam, then diesel, drove the pumps. The old scribbled, sluggish routes of the rivers that choked and spilled across The Fens were diverted, dried up, became ghost rivers, extinct waterways, void of all but their names – a *roddon*, a *slade*. Extinct, but still visible – made ghostlike – from their residues of silt and shell marl winding across the dark and shrinking peat. Shrinking because if you take the water out of peat it is quickly replaced with air, oxidation follows and then bacteria set to work, breaking down the peat so that the plant materials decompose. The impact of this is dramatic: the peat shrinks and the land drops away from under itself, leaving behind the strangest of landscapes, a place where rivers have to be carried like aqueducts above the constantly shrinking fields, where houses are left with their front doors suspended 12 feet above the ground, where trees sit on top of their exposed roots.

And linking the ghost rivers – the roddons and slades – are the outlines of eviscerated lakes or meres, the lakes' white freshwater chalk deposits and marls, like pale birthmarks, still visible against the black peat. Huge lakes were drained and disappeared: Trundle, Ugg, Brick, Ramsey, Benwick, Whittlesea ... all names of lakes that have vanished from The Fens. Whittlesea Mere: the second-largest lake in England, which in the space of two years between 1851 and 1853 went from a place of pleasure yachts and sudden storms to fields of yellow corn.

I watch the female Montagu's harrier circling for a long time with prey in her talons. She seems hesitant about returning to her nest and instead lands several times a good distance away. Then she is up again, circling, the prey still clutched beneath her. Something is unsettling her. I move back deeper into the hedge. Then I notice the farmer has parked his car at the edge of the field and he is slowly walking through the crop, waist-deep in barley. As he walks he picks the occasional barley ear, rubs it against his palm and inspects the grains. They are skittish birds, Montagu's harriers, they live the jittery, nervous life of a ground-nester, often unsettled by the presence of a large animal near the nest site, anything from a passing roe deer to a pheasant. Eventually the farmer leaves the field and the female is quickly down onto the nest.

It is a strange spot for a bird as rare as this (the rarest British breeding bird of prey): a busy habitat, a steady flow of traffic down the road that borders one side of the field. Walkers and horse riders pass along a bridleway which separates the wheat from the barley crop and I notice how the harrier uses this path and the adjacent hedge as one of her linear hunting routes. There is a village two fields away, a glimpse of brick and flint through the trees, though the buildings are barely visible in the shimmering heat. The nest is monitored by the RSPB, and before the field is harvested, the immediate vicinity around the nest is cordoned off and the young harriers fenced in so they do not disperse through the crop when it is being cut.

If a ghost is nothing more than the trace left behind, the scent or spoor of something vanished like a line of snowy shell marl from an extinct river, semi-luminous against the black peat, then the Montagu's harrier is like a ghost over the English fenlands. She is so rare and dwindling a species in these islands, she is barely here at all and exists on the verge of vanishing. She hunts over a landscape that has changed beyond all recognition. A place so drained and straightened that her long-distance linear flight paths could be a symptom of the landscape, a heading out indefinitely in search of a less sanitised – a more disobedient – version of the land.

Huge crowds gathered to witness the draining of Whittlesea Mere. People waded across the wet mud with boards attached to the soles of their shoes, scooping up the thousands of stranded fish into sacks and baskets. The mud was so thick, the boards on their shoes so cumbersome, the fish-gleaners moved like cattle across the lake bed, slow and heavy. Eels shivered and flickered through the remaining pools as if the water was boiling over. So many fish were gathered up that day that several carts were filled and the catch wheeled off to the markets in Birmingham and Manchester. Some months later, after roads and farms had been marked out across the reclaimed land, heavy November rains swelled the rivers causing the dykes to burst, and within hours the lake was a lake once more.

The breached dykes were repaired and reinforced. An Appold pump with a 25-horsepower engine was put to work and Whittlesea Mere was drained again within the space of three weeks. Then came the drying and the cracking of the mud into thousands of fissures and crevasses. Horses were out of the question: even if they were shod with wooden boards the mud was too thick to take their weight, and when it dried the cracks too wide and deep for horses to tread safely. To make the mud pliable the new ground had to be harrowed over and over again by hand. Coleseed and Italian ryegrass were planted first; the ryegrass did better in the wet conditions.

As the mud was worked over it dislodged things – treasures – of startling antiquity: items, lost from storm-havocked boats, that had lain at the bottom of the lake for centuries. Objects caked in mud that when picked up felt heavier than mud and possessed the hint of a bright undercoat, a tincture of light seen through cracks in the mantle. Two of these treasures, a fourteenth-century gilded silver censer and an incense boat, are on display in the Victoria and Albert Museum in London. I went and found them one afternoon in their glass case in a basement room of the museum. Both objects are assumed to have belonged to the Benedictine monastery at Ramsey Abbey situated a few miles from Whittlesea Mere. The censer and incense boat are the only surviving examples of their kind; all similar ecclesiastical treasures were confiscated and melted down during Henry VIII's Dissolution of the Monasteries.

Survivors, gilded rarities, they shimmer under the glass. The deck and hull of the incense boat is worn silver accentuated by a rim of gold leaf around the gunwale. A ram's head at either end, carved into the stern and prow, with curved horns like tiny mollusc shells. The fleece down the rams' necks is plaited drops of silver. The engraving along the gunwale is very delicate like a seam of stitchwork. Next to the boat, the censer is brighter, entirely covered in gold leaf. About the size of a child's face, the censer is designed as a medieval chapter house, complete with miniature lancet windows, battlements, parapets ... The lights in the museum are so bright they make a shadow of the lancet arches in the censer's charcoal-scuffed centre.

As the Montagu's harrier ages there is a gradual uncovering of the bird's eye colour. Both sexes are born with their eyes brown, though their underlying colour, like the mud-coated censer, is a bright jewel-like yellow. Whilst they are still in the nest the males' eyes quickly change to a greyish-brown (a useful indicator in sexing the birds at this age). By their first winter the males' irides have changed again from grey-brown to a clear bright yellow. The brown in the females' eyes fades more slowly, gradually revealing glimpses of the yellow beneath as the brown dissipates. For several years, whilst the brown slowly gives way to the underlying yellow, the female harrier's eyes appear orange. By her third or fourth year her eyes have cleared to the same bright yellow as the male.

For nearly an hour through the heat of the early afternoon I do not see the female. Then at 2.10 p.m. she startles me by flying straight past my hiding place in the hedge – long wings and a whoosh of air. By the time I pick her up again she has reached the end of the hedge. She turns around and begins to fly quickly back down the side of the hedge towards me, twisting, flickering low across the field. As she draws parallel she tilts her head in my direction. Her face seems very small, more like an eye than a face. I can see the dark shading on her cheeks and the white patch around her eyes.

She keeps to the same line, tracking along the side of the hedge. At the point where she passes the dead oak tree she is briefly surrounded by a flurry of white shapes. It looks as if a pillow has burst around her. I refocus my binoculars: the downdraught from her wings is disturbing a cluster of white butterflies. Everywhere butterflies are being shaken up towards her, dusting around the harrier as if she has entered a sudden snow shower.

IX

Peregrine Falcon

Coventry

When I reach the bench I find them as I had left them the week before. Both birds are perched above me, the male on a ledge high on the spire of Holy Trinity, the female on the flèche of the new cathedral, directly opposite her mate. The bench sits in the ruins of Coventry's old cathedral. It is a contained space away from the noise of the city, not unlike the corrie on the side of the mountain in Sutherland, a place of enclosed stillness, where the ruin's walls hold and amplify the sound of the falcons calling to each other.

Coventry is not far from where I live and I was able to visit the peregrines there on several occasions. My frustration with this journey, trying to find the different birds of prey, is that my time with the birds – in each place – was so transitory. I always left reluctantly. I felt acutely the need to spend more time among the birds. So with the peregrines in the centre of Coventry I was fortunate that I could keep returning to them. One June, whenever I had a spare morning, I hurried back to the cathedral ruins, arriving at the bench soon after dawn, the sun turning the cathedral's sandstone a deep chestnut red, flocks of gulls passing over the city in their silent morning height. The falcons were always there, at home, often perched in the same place I had left them.

June: the peregrine's month, when skies are pierced with the male's hunting tracers, when cliffs echo with the begging calls of young falcons who rush towards flight so rapidly they have different names for almost every passing week: an *eyass*, then a *ramage hawk*, then a *brancher*, all before the young have left the nest and learnt to soar.

I used to spend hours as a boy in a secretive glen near our home. Its steep, bracken-covered slopes cut if off from the world; a refuge for roe deer, a waterway for dippers. In the summer months I camped down there amid the reek of wild garlic, swam in the burn's slack elbows.

Bright, cold water: lovely to drink, so cold it burnt your teeth. Halfway up the glen, around a sharp kink in the burn, was the peregrine's cliff. At the foot of the cliff the rock was curtained by a deep bank of moss where dippers hid their nest. The water was so clear I sometimes watched the dippers treading the shallows after the caddis and mayfly nymphs. The current, where it met the dipper's breast, shaped the water like bracken fronds when their young necks are fiddleheads.

Dippers were my water scouts, leading the way up the burn. A sharp *zeet* call, then a short dash upstream to the next boulder where they waited for me, bobbing above the rushing water, their lit breasts, like white aprons, signalling. The dippers lived on the cusp of worlds, half in the water, half out. Jettisoning their buoyancy, they somehow, miraculously, walked along the river bed, foraging there, head first into the fierce currents.

Back then peregrines were scarcer, birds of the peripheries, of upland crags and seabird islands. Their population was still recovering from a dramatic crash in the middle of the twentieth century. From the early 1950s through to the mid-Sixties peregrines had been somewhere awful. Poison seeped into their world: agricultural pesticides trickling down the chain from seed to pigeon to falcon. Organochlorine compounds built up in the tissues of peregrines (also other raptors, notably golden eagles and sparrowhawks). The effect of these compounds was to reduce the supply of calcium carbonate (critical in the egg-formation process) to the egg shells. It caused the shells to thin and grow so fragile that the eggs would crack under the weight of the brooding adult birds. Thieves, looking for eggs patterned with the swirling reds and browns of Mars or Jupiter, found only broken shells. The birds' behaviour addled. Female peregrines brooded empty scrapes or sat for weeks on the wreckage of their eggs. Some peregrines took to ousting kestrels from their nests, hatching and then raising kestrel chicks instead. The population crashed. The birds died like a language, receding to the outer skerries of our world.

They were fiercely territorial, those falcons that came to nest each spring in the glen. Launching themselves from the cliff, tearing down towards me, screaming as they swooped low over my head. I would hurry on up the glen, criss-crossing the burn, then climb one of the steep banks, past the falcons' plucking post, a puff of pigeon feathers turning in the breeze, to where I could watch the birds through binoculars without disturbing them. My memory of the glen is of a world of terrific speeds, the falcons rushing high above me or tearing past my head, the young peregrines chasing after their parents along the cliffs, begging for food. Nothing was ever still in their world. For a few months in spring and early summer the glen was pierced with noise and speed.

The pesticides were banned, the bans enforced: it was not after all too late. Peregrines pulled back from the brink. The population slowly recovered. Old eyries on the cliffs of Devon and Cornwall were tenanted once more. Flocks of pigeons coming in off the Bristol Channel scattered like seed. Rock doves learnt to hug the cliffs rather than expose themselves to the falcon's swoop. As peregrine numbers swelled, the falcons, for so long a bird associated with the margins of these islands, began to colonise urban spaces. Peregrines found sanctuary in our towns and cities, habitats where they were left undisturbed to nest on the tall cliffs of buildings and hunt the large urban populations of feral pigeons. What occurred was an unexpected breaching of worlds, a shift from the depopulated corners of these islands into the centres of population. Peregrines came to live amongst us and their wildness pulsed through our cities.

Listen! Look up! When I hear the falcon I do not hear anything else. The city drops away from under me and there is just the sharp, piercing *kee-errk, kee-errk* cutting through the warm air. The surrounding buildings bounce the sound back so that it rushes at me, amplified, the call rising in pitch, the second note sounding more drawn out. I try to pick out the phonetics of the falcon's call, *ee-ack, ee-ack; kee-errk, kee-errk* ... but it's hard to transcribe and my notebook scrawls with imitations until the page looks like something from a codebook.

From the bench I watch the male – the tiercel – leave his perch and circle out in a wide arc back to the spire on Holy Trinity. A flexing flight, a pirouette: no intent to it. He lands on the spire's weather vane and his landing sets the vane rotating. Every few seconds, as he spins around the circumference of the spire, the tiercel's profile changes, black wings rotating through the mottled white plumage of his chest. The vane keeps on spinning like this, absorbing the force of the bird's landing; for several minutes the whole city lies under the tiercel's rotating gaze.

I get up off the bench and walk around the perimeter of the cathedral ruins along the adjoining streets. In drains and gutters lining the cathedral's walls I find the detritus of falcon meals: parcels of bone stringy with dried sinew, feathers matted with skin, a pigeon's claw with a turquoise ring around its ankle.

Too far inland to be bombed, at the beginning of the war Coventry and the Midlands were felt to be out of range of the Luftwaffe. When France was invaded in the spring of 1940 children who had been sent to stay with relatives on the English south coast were pulled back by their families to Coventry, to the distant safety of the Midlands. When the bombings began to reach the outskirts of the city in June 1940 they did so tentatively like the first wisps of an approaching shower. A ruined house was a novelty, buses did a good trade taking Sunday afternoon sightseers to see the damage. Bomb fragments were much sought after, treasured like pieces from a meteor.

June bends towards July. The light tilts. The air thickens with heat. The bombing raids become more familiar. The falcon young have flown and passed into another stage of their nomenclature: they are *soar hawks* now, learning what they are capable of. Very few of them will survive beyond the winter. On 1 July 1940 the Secretary of State for Air issues the Destruction of Peregrine

Falcons Order. Adults, eyasses, eggs are destroyed up and down the land. The peregrines pose too great a threat to carrier pigeons, lifelines for the military. The order lasts the duration of the war. Cliffs are overseen instead by ravens, the falcon's grudging neighbour.

To reach Coventry, to plot their way accurately, German pathfinders follow a system of radio navigation beams that lead the pilots to their target. A main approach beam is intersected by a series of cross-beams which mark off the decreasing distance to the drop zone. Fifty seconds after the planes pass the final cross-beam an electric circuit on the bomb-release clock closes and the first incendiaries fall over Coventry, lighting up the city for the subsequent waves of bombers.

Peregrines will return to the same nest sites year after year. Signs on the cliffs can help to guide the returning birds: green stains where nutrients in the droppings of last season's young have spilled down the rock and lit it up with algae blooms.

By early autumn 1940 Coventry is increasingly deserted at night. The doors of empty houses are marked *S.O.* in chalk – the wardens do not wish to risk their lives searching for people who are not there. *S.O.* stands for *Sleeps Out*. *Out* means the surrounding countryside, sleeping under hedges, bridges, in barns, burrowing under ricks of hay. Cars parked in pools of darkness. A shilling a head the going rate if you could find a bed at a farm or someone's house, baths were extra. Each evening at rush hour the nightly trek begins, thousands streaming out of the city, prams and wheelbarrows carrying bedding, a long procession of torches glittering in the cold air.

Almost half the city empties out like that each night. The wealthy have places to stay, cars to take them; the poor walk and sleep beneath tarpaulins in the woods. And the countryside at night so vivid and strange: bedding down with the noise of the roe in rut; finding piles of crab apples spilt from hedgerows like pools of green and yellow light; and once the noise of badgers rummaging through a wood, sounding like the wood was murmuring to itself. Hearing the raids coming in, the ground guttural, rumbling. An aftertaste of gasoline in the air. Fires breaking out across the distant city. Then back again come morning, stiff and dewy, the streets smelling of smoke. Dust on everything like a thin layer of snow. Searching for homes and finding just the staircase, the ribcage of the house, leading into thin air.

So that autumn of 1940 there was a constant flow, tidal in its regularity, between the city and the surrounding countryside, a crossing over from one zone to the next, from the dark of the blackout city to the deeper dark of the city's periphery. The city slept outside itself and every morning sleepwalked back into its own skin.

Between the Leisure Centre and the underpass I find him, banking above an empty car park in a tangle of crows. I hear the crows, look up, and there is the tiercel's sickle shape cutting away to the north. All morning I follow him through the city. The old cathedral looms like the great sandstone cliffs of Hoy. I see his reflection swipe the glass front of the museum. Sometimes I lose him behind a building until scolding crows give him away and he appears once more, quick and low over my head.

When he perches, often for long spells, I perch beneath him, keeping watch from a shopping centre, a bench outside an office block, through the window of a bar, open at 8 a.m., *the only place round here you can get Newkie Brown*; the tiercel outside, shifting.

A pub in Birmingham, 14 November 1940. The moon so bright it finds the glasses' sheen. A man in grey overalls bursts in, out of breath. The moonlight seems to clean his clothes. He stands there, gannet-bright. *Just come in the lorry from Coventry*, he says. *You want to go outside and take a look: they're getting it badly tonight.*

The first sign of the raid: dogs barking through the city, hearing the drone of the planes before anyone else. Then incendiaries making swishing sounds like heavy rain. People running into the streets with wet rags to put the fires out. A lady with white hair wears a colander on her head to stop the aircrews spotting her. Bombs coming down like hailstones. Shelters filling up. Candles are lit in upturned plant pots. Blown-off doors are used as stretchers. Hospitals spill over with the wounded. The fires and the moon feeding off each other make it light as day; one boy tells his mother he doesn't like this sort of sun. The air is hot and acrid. Then the electricity and gas are cut. The fire crews have no pressure in the water. Parachute mines look like German airmen bailing out;

people run to apprehend them with sticks, axes, anything to hand. The explosion stops their lungs.

He is hunting now and everything has changed. A falcon at rest creates a truce and the world beneath his perch is suspended for a while. So pigeons can pass casually by, a dipper can glance from rock to rock. But once the peregrine is up, the world that hangs beneath the hunting bird is suddenly charged. Everything is in flight from the falcon. Pigeons sharpen into speed, become almost falcons in themselves: compact shots of speed. Hunted birds have been known to be shocked into tameness, seeking out the safety of people, even allowing themselves to be picked up. Some birds grow so paralysed with fear there are accounts of people walking along a shoreline beneath a hunting peregrine gathering up the cowering snipe, placing them in the warmth of their pockets, each snipe a trembling, unexploded ordnance.

Even a falcon's absence can be felt. The presence of other birds of prey like the kestrel or merlin can sometimes indicate the absence of a peregrine, as if the smaller raptors have moved into a zone of refuge created by the absent falcon. Those peregrines, pesticide-addled, that reared the kestrel young were raising prey. Almost all birds are at risk from peregrines. There are records of geese, black-backed gulls, even buzzards being struck down. MacGillivray was told about the remains of a black grouse found at a peregrine's eyrie on the Bass Rock. The falcon would have had to carry this heavy prey a distance of 3 miles from the mainland. In falconry's heyday peregrines were trained to bring down birds as large as red kites and herons. But prey of this size is really an anomaly; peregrines are essentially a specialist predator in that their preference is for medium-sized avian prey. In these islands, pigeons (specifically rock doves, feral and homing pigeons) are by far the most common source of food for peregrine falcons. Depending on season and habitat, marine birds are also taken, especially black-headed gulls and fulmars. Additionally, corvids, red grouse and waders such as lapwings, golden plover and redshank can be locally important prey species for peregrines.

Some birds, conversely, are pulled towards a falcon's presence. The tiny wren will often nest close to peregrine eyries, gathering feathers from the falcon's plucking posts to line its own nest. Geese, too, have been known to cluster their nests close to peregrine cliffs, benefiting from the falcons' protection against ground predators such as foxes. Gauntlet-runners, the geese choose to walk to and from their nests, rather than risk drawing attention from their hosts. Once, in the glen near home, I took a dog with me and the presence of the dog seemed to ratchet up the falcon's fury. I had a glimpse of how unrelenting and aggressive the birds could be to foxes that strayed too close to their nests. The falcon rushed at the dog (and me), screaming, swooping low over our heads. All I wanted to do was get out of the way.

* * *

Towards eight o'clock the first incendiaries hit the cathedral, singeing the frost from its roof. Four men are on duty to defend it. The fires work through the building, moving from nave to vestry. A large incendiary lands amongst the pews and it takes two buckets of sand to extinguish it. The men rescue what they can: the altar cross and candlesticks, a silver wafer box and snuffer, a wooden crucifix ... They are soon exhausted, soaked in sweat. Steel girders twist in the heat. The organ (which Handel played) is a concentrated blaze. They can hear the drip of molten lead as the roof begins to melt. Throughout it all – throughout the night – the cathedral's tower clock keeps striking the hours. People across the city, when they hear the chimes, presume the cathedral has survived.

Suddenly he is alert, restless, clicking his gaze through every inch of sky. Three pigeons, flying just above the rooftops, seem to slow as they pass beneath him. The tiercel tilts towards them, slips from his perch, throws himself at the pigeons. The city rushes towards him. The pigeons blur then fracture. They drop down into the safety of the street, killing the falcon's space to swoop. The hunt is abandoned and the peregrine is back on his glaring perch.

One hundred and fifty, 180 mph a falcon reaches in its stoop. Waiting-up in the clouds or, when prey is spotted, climbing rapidly to position itself above the target. Then wings are folded back and the peregrine hurls itself at an angle of 30°–45° to the horizontal, sometimes almost vertical. Occasionally aiming wide and then striking from below as the falcon surges up out of the stoop. Often, the force of the impact is such that prey is knocked out of the sky and the peregrine then has ample time to glide down to the crash site. More often than not the falcon misses. If the hunt is not abandoned then the peregrine will beat back up to gain sufficient height to launch another attack, wings folded back, tearing down through the astonished air.

Later that morning he makes a kill. I miss the moment. It happens somewhere out on the edge of the city where I lose him in a fury of speed. When I find him again he is plucking at something on a ledge of the new cathedral. All I can see of the kill is a boil of red flesh. The female stands to one side calling to him, bobbing, agitated. When he has finished feeding he shifts his perch and begins to preen. Feathers drift away from him like thistledown.

Sounding the all clear, crawling out of shelters, mist over the city, a slow drizzle in the cold dawn. Everywhere the crunch of glass underfoot. A butcher's shop still burning, the smell of roasting meat. Houses quiet, missing the whirr of gas and electricity meters. Dust in the food. Kettles filled from rainwater butts. Fires lit. Tea made. Someone serves out treacle tart for breakfast. The outer walls of the cathedral enclose a great pile of smoking rubble. Incredibly, the tower is still intact, blackened down one side from all the smoke.

* * *

Then I am up in their realm. One wing flap and the falcons could reach me. I have climbed to the top of the old cathedral tower, miraculous blitz survivor, 295 feet high, the third-tallest spire in England. There are squirrels carved into the tower's stonework representing the medieval woods that surrounded the city. A slow coiling climb, past bell ropes, their sallies hanging in dusty stillness.

The night of Friday 15 November the Luftwaffe's main bombing force is dispatched to London. Only six or seven planes reach Coventry and drop 7 tonnes of high explosive. It is so trivial compared with the night before that the raid is barely noticed. The roads out of Coventry are streams of refugees wading through piles of glass. The roads have turned to quagmires from all the mud the bombs threw up. Aftershocks: delayed-action bombs, fires reigniting, shattered nerves, screaming when the sirens start again. The night following the raid the city is emptier than at any time for half a century.

From the top of the spire I can look directly across to the tiercel and the falcon still perched on the flèche of the new cathedral. The falcon – the female – is preening. The male has begun his ballet of agitation, flicking his gaze across the city. I am exhilarated by this new perspective, to be among the falcons, at the same height as the birds, looking down their avenues of sky. The female is taller than I had expected. In some artists' depictions of peregrines the birds often seem too tall, their necks elongated as if the perspective is not quite right, but I can see now that once the female stretches out of her hunched perch she unravels a long back. Her tail hangs behind her, tapering to a narrow end. The wrists of her folded wings protrude up to her neckline, making the wing-wrists look like shoulder pads.

Counting the dead. The first mass funeral of the war. Aid pours in to Coventry, donations from across the globe. Four lorryloads of shoes are dispatched from Leicester. In the ruins of the cathedral, on Christmas Day 1940, the Provost broadcasts to the world: *We are trying, hard as it may be, to banish all thoughts of revenge.*

The male flexes his wings, peers over his ledge and drops. From the tower I can keep track of him more easily. I watch him gaining height, circling over the city. He moves towards the north through a vast acreage of sky. I can see for 20 miles in every direction. I watch him hanging above the city and for long periods he is just a black speck against the cloud. I follow him as he starts to track to the east and suddenly I am aware he has dipped into a stoop, my binoculars trembling to keep up with his change of speed. The next ten minutes I spend trying to locate him over the city's outskirts. I trace my binoculars over every corner of sky and pause to unpick each speck and dot against the blue. But the heat blurs things and I have lost him again in the rising haze of the day.

I walk around the roof of the tower still trying to spot the male. My feet crunch the claws and bones of pigeons that have dropped from the falcons' plucking post above. I find a jackdaw's head lying on the roof. Dried, mummified in the heat, not yet a skull. I crouch down to take a closer look. *One for sorrow, two for joy …* The rhyme starts before I realise my mistake: a magpie's head, a seam of metallic green running through the black feathers.

I cannot find the tiercel. He is shimmering somewhere above the city. I walk around the roof of the tower and pause opposite the falcon still on her perch on the new cathedral. I can see the yellow orbit of her eye, the black sunspot in the centre of it. Then she lifts, turns, and flies towards me.

X

Red Kite

Upper Tywi Valley, Wales

MacGillivray reaches Manchester on 13 October, the distance he has come starting to resemble a migration. Six hundred and fifty-six miles since he left Aberdeen thirty-seven days ago, a steady bearing southwards, picking up momentum, barely stopping now, as if London were all downhill from here. Weaving through the streets of Manchester, trying to find lodgings for the night, but everywhere refusing him. They only need to open the door a chink and take one look at his mud-splattered coat, his worn-out shoes, his grass-stained knapsack ... It is not that they doubt he can pay, rather, that if they let him enter, they fear they would be letting in the weather.

Manchester is a low point. The miles are starting to tell. MacGillivray is exhausted and cannot shake off the feeling he is being followed through the streets. It is strange he should feel so haunted here amongst the cluttered city when the loneliness of the moor has never troubled him. He is trying to persuade a landlady to give him lodging for the night; she is wavering in the doorway and all MacGillivray wants to do is find a place to rest. He would like to curl up there and then and I wonder, if he could find a quiet spot away from the crowds, the coal wagons and beer wagons, could he not make a nest from all the plant cuttings he has in his knapsack and bed down instead in them for the night.

There is a species of goshawk, the beautifully named Dark Chanting Goshawk, that builds its nest by selecting twigs with social spiders living on them. The spiders quickly consume the hawk's nest with their silk so that the structure becomes hidden – concealed from predators – by a bulb of silk. If MacGillivray could build a nest like that, which hid him, protected him, he could forgo the lodging house and sleep instead amongst the noisy streets of Manchester, despite the sewer stench, the stink from all the tanneries, the butcher stalls and tallow chandleries.

There is no doubt now, there is definitely someone following him. He has had enough of this and so he whirls around to face his stalker:

– Sir (*the man says to MacGillivray, as if he had spoken to him*).

MacGillivray says nothing.

– Do you know me?
– No, I do not.
– Upon my word, I beg your pardon. I thought your name had been John Parkins.

I love these tetchy conversations MacGillivray has along the way, these brusque rebuttals. It seems that anyone who spoils his momentum is unwelcome. Pity the farmer, for instance, he met on the road outside Lancaster who was foolish enough to remark that MacGillivray seemed in a great hurry …

– What? (*MacGillivray answers*).
– You are in a great hurry.
– No I am not.
– Aren't you?
– No.
– Where have you come from?
– From the North.
– Is there any quarrelling there? Any fighting?
– No.

– From what part of the North have you come from?
– From Scotland.
– Where are you going?
– To London.

MacGillivray becomes so irritable he almost punches him. Poor man, he starts to back away:

– Farewell, farewell, good morrow, good morrow …

The only exception to these prickly exchanges I can find was the poor man, *turned off by his laird*, that MacGillivray passed on the road outside Elgin when he was walking home one time to Harris from Aberdeen. MacGillivray stopped and gave the man a shilling and apologised it was not in his power to give more, and the man blessed MacGillivray and said he had given too much.

MacGillivray leaves Manchester at ten o'clock next morning. When he reaches the outskirts he joins the London road and settles, more relaxed now, into his stride. Behind him the city's factory chimneys look like colonnades of rain. What did the farmer he barked at mean by *any quarrelling, any fighting?* MacGillivray is leaving Manchester just a few weeks after the Peterloo Massacre, and though he does not mention this in his journal, he must have been aware of what happened, that the region felt charged and changed and fragile. What with his experience of these forces at work in the

Hebrides, of the wholesale displacement of people, he must have recognised what happened at Peterloo – the charge of the cavalry troop into the crowd – as another, more brutal branch of this work, the clearance of people who are in the way.

He passes a milestone – *To London 182 miles* – and his spirits begin to lift. For the first time, London – the British Museum with its great collection of birds – starts to feel in reach. He walks through Stockport and on through Cheshire, approaching the green hills of Derbyshire. His pace picks up, he even (whisper it) starts to feel quite cheerful. A quick calculation: yesterday he covered 24 miles, London is now 158 miles off. Today is Friday. MacGillivray proposes to be in London by two o'clock on Thursday. That's 26 miles a day on a budget of one and twenty pence per day. He would like (of course he would!) to travel more cheaply, but why torment himself with bad meals and scratchy beds when there is no need for that. After all, he has nothing to prove: he is confident of his ability to walk for miles without food. He once travelled 240 miles on just twelve shillings, but there was also the occasion when he spent (he shudders to recall the lack of prudence) fifteen shillings in a day! These are his extremes.

Red Kite

The mouth is wide, measuring an inch and
two-twelfths across ... the oesophagus six inches and
a half long, the crop two inches in width; the stomach
round, and two inches in diameter, its muscular coat

> very thin. The intestine five feet long, from four to two and a half twelfths in width, until the commencement of the rectum, which is half an inch wide, and forms a large globular dilatation ... Wings extremely long, broad, narrow, but rounded at the end; the third quill longest, the fourth almost equal, the first short; the primary quills of moderate strength, broad, toward the end tapering, in-curved, with the tip rounded, the outer five having the inner web cut out ... Tail very long, broad, forked, or emarginated, of twelve broad feathers ... The flight of this bird is remarkably elegant, the lightness of its body, and the proportionally great extent of the wings and tail, producing a buoyancy which reminds one of the mode of flying of the Gulls and Jagers ...

I am sitting on a wall at the far end of a car park in Llandovery, strapping the tent to my rucksack. Late February, a granite light, the sky looks like it has been washed with silt. A pair of red kites come in low over the town from the north. They circle above the car park, one of the birds calling a high, echoing whistle. Not unlike a buzzard, but the kite's call sounds higher, quicker, less resonant.

I drive north out of the town following the river Tywi into the hills. Higher up the valley the land is under mist. Blocks of conifers are darker shapes inside the mist. There are glimpses of oak woods on the steep banks above the river; birch trees in amongst them lighten the oaks' colour. I have to slow the car to crawling pace, put the fog lights on and creep along the narrow road. Through a village smoking in the mist, then a bridge across the river and the road is suddenly even narrower. I park up, wedging the car out of the way under a tall bank. Tree roots, crimson ligaments, show through the turf where the bank has been cut back. Then the familiar ritual, feet up on the car's rear bumper, boots loosened and threaded, backpack tightened.

If the Montagu's harrier is all buoyancy, lightness and drift, then the red kite is pure agility, pure manoeuvrability. And if the Montagu's harrier is all wing, then the red kite is all tail. MacGillivray wrote, *It is his tail that seems to direct all his evolutions, and he moves it continually.* In the Upper Tywi valley the kite is known as *Boda wennol* (the swallow-tailed hawk). Further south in Wales, she is *Hebog cwt-fforchog* (the falcon with the forked tail).

Everything in the kite's design lends itself to agility: long-winged, light-framed, with a tail that is shaped to decrease drag and increase lift, a tail that can perform miracles inside the air. The 'swallow-tailed hawk' has swallow-like agility. A kite can hold itself inside the bumpiest of thermals, constantly adjusting its tail to keep itself in place to scrutinise the ground below. And a kite can suddenly turn on a sixpence, despite its great size, in order to twist around and take a closer look at something it might have missed. What is so mesmerising about the red kite is that a bird of its size – a bird with such long wings – can perform such aerial acrobatics.

You would think that it was too gangly for such aerial dexterity. The kite's size constantly belies its behaviour in the air.

The forked tail enhances manoeuvrability, it gives the bird stability through the tightest of turns. The span of the fork can be varied depending on how wide or tight the turn. The sharper the turn, the wider the fork, and vice versa. It is no good, such a forked tail, in a densely packed habitat like a wood, as the tail's outer primaries – the long exposed sides of the tail – are vulnerable to damage through snagging branches. So kites prefer to nest in roomy deciduous woods with ample space to reach their nests. The sparrowhawk, in comparison, has a long straight tail which is less vulnerable to damage as it hunts at speed through the clutter of a wood.

What the kite needs – like the swallow, like the harrier – is open space to range and forage over. Woodland is used by the kite only for nesting and roosting. The rest of its time is spent hunting over open landscapes. Of all the birds of prey the kite is the one that lives the most inside the air. She does not belong on the ground, does not make sense unless she is in the air.

I pass through a quiet farmyard. Nobody about, not a soul since I left the village. But I notice, as I brush close to it, the warmth coming off the bonnet of a tractor. I follow a quad-bike track out of the farm, spilled wisps of hay in the mud. A path filters off the track and starts to climb between gorse and hawthorn, across beds of flattened bracken. This is the kite's backyard, the *ffridd*, the tangled vole-rich lower slopes, equivalent to the margins between the moor and fields the hen harriers frequent over Orkney. Unkempt no-man's-land, so critical to foraging birds of prey.

The path takes me through a small wood, dripping and still. A kite turns out of the mist and flickers slowly over me. I have seen several kites already including the ones over Llandovery, but every kite is a gift because its flight is rarely hurried and you have time to follow it, watch what the bird's gymnastics do to the air. I think how different the experience is to following the merlins out on the flows of Sutherland, those tiny balls of energy fizzing out across all that distance. The kite is the least linear of raptors, it spends its time unravelling imaginary balls of string in the air.

I walk for several hours into the hills. The mist accentuates the quiet. I find fox prints in the mud, and at one point, the mist stirs enough for me to make out three chequered black and white wild ponies grazing on the hillside. An hour before dusk, I make camp in a small oak wood hanging above the river. I pitch my tent on a narrow clearing in the bracken. As I work a kite takes off from one of the trees in front of me and glides across the river. An hour before dark the mist lifts and I can see clear down the valley. It is clearer now, at dusk, than it has been at any time during the day. The kite is still in view and I watch the bird working her way down the valley, backlit by a thin layer of blue.

* * *

Each kite is a gift because more than any other bird of prey they are indulgent of you. They don't mind you being close to them and, outside the nesting period, are little troubled by human presence. One afternoon I followed a red kite foraging over the ground between a busy ring road and the outskirts of a town. The kite banked above houses, hunting over side streets and alleyways. At one point I watched it hovering just above me and saw it drop something, a vole I think. But before the vole could reach the ground the kite swooped down after it and retrieved it in mid-air. What I remember most from the incident was not the showy acrobatics, more the bird's dramatic turn of speed. All afternoon it had glanced over rooftops and hedgerows so slowly it was fairly easy for me to keep pace with it. But when the vole was dropped, the kite paused for a split second then burst after it in a sudden flex of speed.

They are more tolerant, red kites, of human beings and human spaces than other birds of prey. Kites are essentially scavengers, and like other scavenging raptors, such as vultures, they are often drawn in large numbers to the rubbish produced by humans. Some Asian cities have large resident populations of black kites, and English cities too were once home to substantial numbers of red kites, protected by royal statute because of the important role the birds performed in gleaning the filthy medieval streets. Medieval London was famous for its red kites and the birds so brazen in their interaction with people it was not unknown for a kite to snatch bread from a child in the street, or a cap from off a man's head to line its nest with.

Piratical; fanatical nest decorators: it is not improbable that London's medieval kites were as bold as this. Kites are well known for their piratical habits of harassing corvids and other raptors to drop their prey, utilising their aerial agility to harry the other bird into spilling its catch. Then it is the simplest thing for the kite to swoop down and pluck the tumbling item out of the air. Nests are adorned with anything that comes to hand: caps, handkerchiefs, plastic bags, discarded lottery tickets ...

In Wales I found red kites over car parks, playing fields, allotments, as well as across the Cambrian mountain heights. There seems to be no other bird of prey that crosses over so fluidly, so constantly, into our world. The red kite is a bird that lives in the slipstream of human beings, gleaning, foraging around us. They will follow ploughs and harvesters and take scraps left out in people's gardens. They have even been known to tear net curtains from open windows to decorate their nests.

I sit outside the tent as it grows dark. Cold air has sunk into the valley and my breath is suddenly visible. Moths are out, flickering over my hands, whirring past my ear. There is a scuffling in one of the trees, something scratching at the bark, then a red squirrel scurries down the trunk and hops away over the crunchy leaf litter. If I had not seen it first I might have thought a larger animal, a deer or sheep perhaps, was moving through the spinney; in its frenetic scrabbling, a squirrel can make so much noise inside a wood.

All afternoon kites had come down out of the mist like sudden angels. As I walked up the valley I kept putting up a buzzard that flapped heavily away from me then waited round the next bend before I came into sight again. The buzzard seemed huge and slow in comparison with the flickering, twisting kites. Halfway into the hills a raven flew close and low, its glossy blackness lit up against the winter grass. There was frogspawn all over the hillside, stranded on clumps of reeds, as if it had drifted there like blown sea spume.

The light going now and at the last, before I turn in for the night, a buzzard rushes into the spinney, just a few feet from where I am sitting. It swerves between the trees and I think, because it is flying so fast, it must be hunting. But then it lifts out of its flight and lands on a wide branch. There is very little light now but I can just make out the dark shape of the buzzard, settling its weight on its perch, preparing to roost.

I am up before it is light. I did not sleep much. There was a small stream a few feet from the tent and all night, with my ear pressed against the ground, I kept waking to the sound of water, sounding as if it was running under me. As I sit outside the tent at dawn, making tea, there is a movement in the bracken. Stepping out into the clearing, right in front of me: a fox, white-chested, a brightness to her coat, made brighter – redder – by the rusty bracken behind her. Then she turns, coils away from me, and begins to climb the steep bank. After a minute she stops, glances back towards me. She repeats this several times, pausing, turning her head to look in my direction, finally disappearing into the bracken.

Take a hammer or an axe, whatever is to hand, swing it round and round you in a circle by its handle and, at the point of greatest energy, or before the dizziness overwhelms you, let go the handle and let the hammer fly. Retrieve it, and when you do so, mark the spot where it has landed. Then walk back to where you started and repeat the action. But this time send the hammer off in the opposite direction, until you have flung it out to every compass bearing. Then draw a line between all the spots where the hammer landed. In this way you can mark out the perimeter of your territory, your plot, your home.

This is how squatters in the first half of the nineteenth century would claim a patch of ground for their own on the marginal lands, the *ffridd*, of the Upper Tywi and elsewhere in rural Wales. The point where you swung the hammer had to be the front door of your house. But the house (and herein lay the catch) must be built in a single night in order for the squatter to validate their claim on the land.

Ty un nos: a one-night house. The key was to prepare the roof in advance and gather as many friends as possible to help with the night's work. The walls and roof had to be up and, crucially, a fire lit in the hearth so that smoke could be seen drawing from the chimney at dawn. Surprise: a house!

I looked for traces of these houses while I walked through the hills of the Upper Tywi, drawn to poking about ruins marked on the map. In most cases I found very little, just the skeletal outlines of buildings, soothed over with mosses, riddled with bracken. These one-night houses in Wales coincided with the advance of Parliamentary Enclosures in the first half of the nineteenth century. Access to common lands and upland pastures was rescinded by landlords, forcing the poor to try and make a living from the scrappy wastelands around the margins.

A clearance is an ongoing process. It only succeeds in displacing a people to somewhere they will be in the way again. I found this with the story of the people shunted around the Morvern peninsula until they ended up dumped on the barren outcrop of Oronsay. But even there, of all the godforsaken places, they were not allowed to settle long before they were picked up and herded on. A common name for the Welsh one-night house was 'labour in vain'. More often than not the squatters' cottages were destroyed by farmers who felt their own grazings were being encroached. Fences, hedges, those hammer-marked perimeters, were dismantled and often just as quickly put back up again by the squatters. Disputes rumbled on like this for years. Sometimes squatters' houses were destroyed decades after they had been constructed when commissioners judged that the land they were on should be enclosed. You try to establish a toehold on the land, try to cling on to it, but that hold is just as soon pulled from under you.

* * *

In the first half of the twentieth century the hills and hanging woods of the Upper Tywi became the red kite's last refuge in these islands. A tiny, remnant population, dwindling, clinging on. At their lowest ebb, in the 1930s and early 1940s, there were perhaps no more than ten pairs. They entered a genetic bottleneck and for a long time afterwards all Welsh kites could trace their ancestry to a single female bird. That was how close the birds came to extinction in the British Isles.

Despite its remoteness, its largely un-keepered woods and hills, the Upper Tywi was never a safe refuge for kites. It was just where they ended up after they had been driven out from everywhere else. But the bird's status there was always precarious. Egg collectors, the limited gene pool, the climate of those rain-drenched, infertile hills: there were so many ways for the kites to fail in Wales. And for decades the few remaining birds lived a miserable banishment, their toehold on the land constantly pulled from under them. Even some of those responsible for safeguarding the few nests were corruptible and eggs were frequently stolen to order. When the Welsh kite population did finally begin to increase in the 1960s it was not because more birds were being born, rather that fewer adult kites were dying.

Britain's commonest bird of prey reduced in the blink of an eye to a rump, to just a handful of birds. The kite's undoing facilitated – accelerated – by that tolerance of human beings and human spaces. A large, slow bird swimming leisurely overhead made the easiest of targets; its propensity for carrion made it just as easy to trap or poison. We preferred (we still prefer in some cases) our birds of prey to be banished to the margins. We drove them to the furthest reaches, to the outliers of these islands, like the last lone sea eagle living out her days on Shetland. The Upper Tywi was the red kite's Shetland, its Oronsay.

But when we let them back in, as we have recently with the red kite through successful reintroduction schemes in England and in Scotland, their presence is transformative. The birds transform – they restore – the land. A landscape without raptors is an unnatural one.

People were found to be in the way once more in the Upper Tywi valley in the mid-twentieth century. The first the farmers of the area heard about it was a summons to a public meeting at Llandovery Town Hall on 18 November 1949. There they were told, abruptly, dismissively (in English), of the Forestry Commission's scheme to plant the largest forest in Wales, covering 20,000 acres, swamping forty-eight farms. The forest would take forty years to come to fruition. Farmers could volunteer to sell up or, failing that, the Commission would have no alternative but to issue compulsory purchase orders.

So the 'Battle of the Tywi' began and the farming community, angered by the manner and tone with which the Forestry Commission had proposed the scheme, organised itself to resist and petition against the purchase order. The 'battle' rumbled on for several years and ended with this statement in the House of Commons on 31 January 1952:

– Mr Baldwin asked the Minister of Agriculture whether he can yet make a statement about the proposals for the compulsory purchase for afforestation of a large area in the Towy Valley.

– Sir T. Dugdale: Yes. I should welcome an opportunity of explaining the position. A draft Order for the compulsory purchase of some 11,450 acres was published on 18th September, 1950, objections were lodged and a public local inquiry into them was held. The report of the inspector appointed to hold the inquiry was submitted to me on 20th November, 1951.

I have examined the proposals and considered the objections as reported by the inspector. After considering all the circumstances, including the country's financial and economic situation, my conclusion – with which the Forestry Commission concurs – is that it is not expedient to proceed with the project, which would involve heavy capital expenditure as a preliminary to a programme extending over 20 years. I have, therefore, after consulting my right hon. and learned Friend the Home Secretary and Minister for Welsh Affairs decided not to make an Order.

I sit down in a small field just above the river. For a while I watch a pair of ravens, vocal, busy about their nest site on a cliff high above the valley. Then a kite glances above the ridge. The rest of the morning I spend sitting at the edge of the field following the kite as it hunts slowly over the adjacent hillside. Behind me, great tits call and skitter through the hanging oak wood. Sessile oaks: hobbled and mossy, ferns growing along their thick shoulders. Sheep are inside the wood. Four skittish ewes with winter coughs, coughing at me when I disturb them coming down through the trees.

Sometimes the kite drifts towards me and hangs above the river. Its tail tips a fraction, pitching the bird into another angle of drift. The tail is almost fishlike in its flickering, waving movements. Other kites pass across the valley and I try, unsuccessfully, to distinguish the sexes. I read that the male is a fraction smaller, a fraction lighter, more agile, his tail working, flexing constantly. The female's tail is not as deeply forked, her wings are slightly longer, broader, more pointed at the tips. Later that day I watch a pair of kites flying close together, mirroring each other, one kite riding 8 feet above the other, a synchrony, a courtship in the air.

Throughout the morning buzzards and ravens interject into the kite's airspace. The kite amongst its neighbours is so much slower and less purposeful; even the buzzards are usually heading somewhere, crossing from left to right, or working to gain height. The kite, it seems, is not going anywhere. It is too intent on scrutinising the ground, rotating, adrift in its own gyre.

During the nesting season the relationship between the three species – kite, buzzard and raven – is fraught with squabbles. Red kite nests situated close to breeding ravens will often fail. If you hear the unmistakable quickfire coughing of a raven's alarm call it is almost certainly aimed at a kite or buzzard passing through the raven's territory. Mid-air skirmishes, dive-bombings, are commonplace at this time of year and the cliffs ricochet with the ruckus.

Despite its size, the kite is not a powerful bird of prey. It relies on the stronger-beaked buzzard or raven to puncture a new sheep casualty. So the dynamic between the three birds can shift as the kite waits its turn behind buzzard and raven in the pecking order at a carcass, with the usual attendants – carrion crow and magpie – darting in between to snatch what they can. There is also a staggering of laying and hatching dates between the three species, as there is between the harrier species and between other raptors that share a habitat. Red kites breed seven to ten days earlier than buzzards, enabling the kites to feed their young on the potential glut of young black-headed gulls, corvids and woodpigeons; buzzards time their nests a little later to coincide with the abundance in young rabbits and voles.

Kites also take voles, immature rabbits, leverets, though rarely anything larger than a small rabbit. Live prey is usually dispatched with the bill, rather than the kite's relatively weak talons. Field voles, as with so many British raptors, can be a locally significant prey and fluctuations in the vole population can impact on the kite's breeding success. Red kites in Wales bred poorly in the years when myxomatosis in rabbits was at its peak. But no raptor is more of a generalist, more adaptable, more omnivorous, than the kite. It seems to have a characteristic taken from almost every other bird of prey I've met.

As much a scavenger of carrion as the sea eagle, it will also forage for worms alongside buzzards on moist fields in the early mornings, or hawk for dragonflies and crane-flies like a hobby. It will eat anything that is dead or nearly dead. Lambs' tails and lambs' scrota are a favourite titbit in Wales, even the rubber rings used to dock them are consumed. Fish, reptiles, amphibians are also taken across the kite's European range. A nineteenth-century egg collector in Wales used to warn people climbing up to kites' nests to *Beware of half-killed adders!*

Communal winter roosts can attract spectacular numbers of red kites. So many birds flocking together to spend the night you wonder how the trees can bear their weight. MacGillivray relates an incident at a kite roost of fog suddenly freezing one night and affixing the birds' feet to the boughs of a tree. In the morning some boys clambered up the tree and retrieved fifteen red kites, prising the birds from the branch as if they were dislodging a row of icicles.

More gregarious, more relaxed about territorial spacing than most other raptors, kites can concentrate in very high densities. Their only territory is the immediate vicinity around the nest. Everything else is air! Adult birds are also prone to returning to their natal nesting areas to breed. So there are pockets of the country now where you can walk all day through a swirl of red kites. The birds function as a community: red kites are chiefly scavenger birds, foraging in loose groups, 'network foraging', relying on each other to pass the word along when a carcass has been spotted, much as vultures do. But this tendency to bunch together, to concentrate their numbers at high densities, also exposes the kite to harm. Still today, when carrion is laced with poison and deliberately planted in a red kite area, there follows a wholesale destruction of the birds. Other scavengers – buzzards, corvids, foxes – will be casualties too, but kites invariably pay a heavy price.

After the compulsory purchase order was dropped, the Forestry Commission employed softer tactics in the Upper Tywi. This time they approached farmers individually, using Welsh speakers as intermediaries. Many of the hill farms were still reeling from the losses to their flocks during the harsh winter of 1947, and once one had been persuaded to sell, many others followed suit. In the end more land in the Upper Tywi was afforested than had originally been proposed. Farms were cleared, even the oak woods were cleared to make way for the conifers. The stubs of one-night houses were covered over, the kites' foraging grounds covered over.

In the afternoon I return to the car park in Llandovery. I climb the small hill in the rain and stand amongst the castle ruins. Jackdaws spill, squabbling and chattering, from the town hall roof. Then two kites come into view, glancing over chimneys and television aerials. I watch as they drift across the town, unhurried, checking every yard and garden. One of the kites adjusts its path and starts to glide towards me. I can see its ash-grey face and nape, the white markings on its underwings, the flash of red between its tail and breast. Then the kite ducks out of sight behind one of the castle walls and I think I have lost it, but next moment it is suddenly above me, holding itself in the rain just 10 feet from my head. Face tipped downwards, beak angled towards the ground. Its eyes follow the line of the beak. Its beak opens, as if it is going to call.

XI

Marsh Harrier

Isle of Sheppey

I go down to the north Kent coast in January. It is snowing and there are thin drifts across the motorway. Early morning, still dark, the traffic moving to the outside lane where the drift is thinner. Flashing lights in the rear-view mirror, then a brief clatter as my car is brushed with the discharge from a gritter. Single file, down to 20 mph, headlights seem to make the snow fall slower.

That day the Isle of Sheppey felt abandoned to the birds, utterly given over to them. The marshes under ice, the ditches frozen, snow banked up on the windward side of dykes. I spent the day from dawn to dusk walking over the island's eastern marshes in an ice-shot wind beside a frantic sea. That wind never let up, never moved on; it had snagged itself on the island and spent the day trying to tear itself free. All day snow clouds waited off the coast, banked above the brown sea, waves of snow rolling in over the mudbanks and brittle reed beds. But when the snow hit the wind took it and flung it out across the marshes so that it was nothing like snow, had none of snow's quietness. It was snow corrupted into ice, snow consumed by wind. It made the day impossible, a madcap wind that yelled and spat at me relentlessly: *Keep your gloves on*; *face away from me*; *stop dilly-dallying* … And I found that even taking my gloves off for a few seconds was so painful I gave up on trying to write notes or photograph. What was there to photograph? The land speeded up in a blur of snow and wind and snowflakes smudging the camera's lens.

But birds were there on Sheppey in their thousands, huge flocks of woodpigeon, lapwing, curlew and starling. And none of the flocks, it seemed, were deterred by the wind. At least, they had not been grounded by it and were buoyed and buffeted and flung everywhere inside the wind's madness. A small covey of red partridges split from under me and the lunatic wind grabbed the birds and hurled them across a ploughed field. The partridges whirled away so fast, I thought: they will not be able to stop, unless they can turn into the wind they will be flung across the marshes, over the mudbanks and out to sea. And if they reach the sea I knew I'd lose sight of the birds, not because of the distance (the sea was only two fields away) and not because the waves might grab the birds, but because the sea was so churned with mud and sand it had lost its usual grey and turned a muddy brown, so that the partridges, when they cleared the last field, would simply blend into the ferrous backdrop of the sea.

And in between the restless flocks – between the passerines, the waders and the ducks – were the flocks' magnets, the avian specialists: merlin, sparrowhawk and peregrine.

If I had to fold one of the places on my journey onto another, it would have to be Orkney folded onto – twinned with – Sheppey. Not so much for the wind or the sting of the sea's breath on everything, nor for the fact that if Sheppey broke its moorings and floated out of the Thames estuary, it would be blown north like the dregs of the Armada until it crashed into Orkney and wedged itself in one of the steep-sided geos on South Ronaldsay. Rather, for the sense that birds had overrun – had inherited – both island landscapes. Also, a sense that I had caught up with many of the birds again from Orkney here in their winter quarters on the north Kent marshes: curlew, lapwing, hen harrier and merlin.

I am crouching in the lee of a sea wall eating a freezing lunch (my water bottle has turned to mushy ice), trying to ignore the wind's harassment: *Go away wind*; *leave me alone*; *I'm trying to eat a sandwich*; *look what you have done to my fingers … they have turned a colour I do not recognise* … Then, above the pale reed bed, a sickle-shape echo from Orkney, from the Flow Country, from the mountains of Lewis and Harris. My first thought: a peregrine. But, no: too small, too low, skimming far too low across the saltmarsh. Not peregrine, but merlin.

Of all the raptors the merlin is the blink-and-you'll-miss-him bird. And in the seconds it takes me to gather my things, pack up my lunch, he is gone. This merlin was so charged with wind he could have been flung as far as Essex or Orkney for all I knew. But I find him again high above the marsh and it is as if he clears a space around him, as if the sky is purged of all other birds until there is just the merlin, an erratic black dot – an eye spot – backlit by the snow clouds. Kinetic, wind-invigorated, he rushes across the grey sky then drops suddenly down, flicking small birds out of the reeds, veering after them. I follow him across the marsh, thinking: is he hunting, what is he doing? He seemed to be testing the birds, looking for the stragglers, the slow-to-react. The reed birds spat out from under the merlin but he did not latch on to a pursuit, he just ploughed on, flushing out the tiny birds, broadcasting them like seed. The last time I see him he is perched on a bank of snow, catching his breath, glaring back at all the dislodged birds in his wake.

* * *

If a day can give up on itself long before it is over, even before, for goodness' sake, it has reached its midpoint, that winter's day on Sheppey was such a day. But then the day never really stood a chance faced with all that wind and snow. It was better that it closed itself down and let the onslaught continue through the dark. So it was strange to see figures coming towards me across the marsh just at the point, in the early afternoon, when the light was being raked out of the sky and the day was perishing. Wildfowlers: coming out of the grey and the snow with their dogs, faces buried under balaclavas, bulky camouflage jackets, shotguns sleeved over shoulders, nodding as they passed me. Strange apparitions, they were only briefly real. Though I knew they were not spectral because, at the moment we passed each other, I heard them cussing their dogs to keep back.

The first marsh harriers I met on Sheppey were a pair. Huge dark brown birds, untroubled by the wind, working a small field that bordered the marsh. I stood above the harriers, looking down on them from a slight hill. They shifted through the wind lifting over hedges and fences, peering into field margins, scanning the dead winter grasses. More than their size, their slow flight made them conspicuous. They were using the wind to hold themselves up, turning in to the wind to slow and steady their flight. The wind was so strong, at times it held the harriers static, pinned them in the air like giant kestrels. They were a female and male. Too distant to make out the patterns of their plumage but the male lighter, greyer; the female, when she came close to the male, noticeably larger, heavier.

I was so pleased to see them. There are some days when I spend all day searching and find nothing and that's as it should be. But it is such a relief when I do catch up with the birds. I was worried that the day's weather would have grounded any harriers who had stayed on Sheppey for the winter, but they were unmistakably harriers, could not have been anything else, gliding on V-shaped wings, sliding over the ground, foraging, occupying the harrier zone which is theirs and theirs alone, those first 2 to 6 metres above the ground, within earshot of the slightest squeak or rustle. Sound-gatherers, radars, listening in to the undergrowth as no other bird of prey, except for those great auditory hunters, the owls, can do.

I walked down the hill towards the harriers, hoping to get a closer look at the birds' colours. The male marsh harrier does not have the pearl-grey colouring of the male hen and Montagu's harriers. Instead the male marsh's plumage is a distinctive tricolour, black wing tips, light grey wings, a chestnut undercarriage. The contrast between these colour bands is like early morning fireplace ash before it is disturbed, the undercoat of grey, the black charcoal splints, the red fibrous imprint of the burnt-out logs.

I stalked the two harriers under the cover of a hedge which fell down towards the marsh like a slipway. But when I reached the field where they had been hunting the harriers had split apart and were a long way off, heading east, flapping slowly along the high-water mark.

Even within that harrier zone there are demarcations – holding patterns – for each of the harrier species. The marsh harrier tends to fly a little higher than its congeners, the hen and Montagu's harrier, relying more on its eyesight than the other species to peer down into the tall reed beds that make up so much of its hunting range. The hen harrier (the greater vole specialist) hunts perhaps the lowest of the three, scything the ground. The hen harrier also has the most pronounced facial ruff of the three resident British harriers. And as the ruff is the harrier's radar rim, of the three species, the hen harrier has evolved the most efficient sound detector to pick up the minute patter of a vole moving through its grassy tunnels.

But the marsh harrier is as much a sound detector, as much a listener-in as its harrier cousins. And the panicked splashing of a duck with her flotilla of ducklings through the reeds must flood the marsh harrier's hearing. At night, roosting on the ground, often in open spaces, harriers depend on that hearing to listen out for threat. Roosting sites are selected for their acoustic properties: reed beds, barley crops, dried mudbeds, all places that crackle and ripple when entered by a predator. A bird that can detect the sound a locust makes feeding along a branch must also hear the footsteps of a fox.

The harrier's face is a scoop, a shallow drinking cup, an Elizabethan face, rimmed by the silky, light-reflecting ruff. The ruff itself is like a thick plait, a ring of closely packed barbules layered around a rim of skin. The ruff can inflate, puff out, increasing the surface area of the face, increasing the harrier's ability to capture sound. The face is really a giant ear. It works like an ear, or, rather, serves the ears, scooping up sounds and channelling them into the large ear openings set behind the eyes. The harrier's facial disc is always working, drinking up sounds, weighing them, tasting them in the feathers' nerves, reflecting them back off the ruff, through the ear-coverts, into the large ears.

I walk north-east across the saltmarsh, following behind the two harriers. In the distance: a hamlet crouched low on the headland, hunkered in on itself. It is slow going against the headwind and I am amazed any bird could beat into such a wind. Then I reach the island's east coast and there is the sea again, brown and unrecognisable. I walk up over the dunes and peer down at the beach. Between the narrow road and the dunes is the most incongruous sign, as bizarre as my ostrich encounter in the fog above Bolton: *This Stretch of the Beach is Dedicated to Naturist Bathers*. The thought of being able to take your clothes off, let alone bathe in such a place ... It is impossible to imagine the beach in light and warmth. There is no shelter, even among the dunes, and the beach is all crashing, roiling noise, spume-flecked, wind-wrecked, abandoned to the gulls and wading birds. Past the dunes: a bric-a-brac of summer beach huts and hauled-up, upturned boats, the colour gone out of them, everything shut down and weighted down by stones and ropes. A place in such deep hibernation you could kick and yell at it and it would not stir. In front of the huts the beach is demarcated with wooden groynes, the space between each of them filled with wind and shrieking gulls.

I turn at last inland. A brief dose of wind-relief as I shelter behind a caravan on its breezeblock stilts. I watch a flock of lapwings shivering over the fields. Then, glancing past the lapwings, closer to me, a marsh harrier: a male, working the seam of a ditch, gliding low over an abandoned saltworks, flying into the wind to slow himself down.

The harrier hunts through an interplay of sight and sound. The harrier's ears are not positioned asymmetrically (like an owl's), so it does not possess the owl's supreme ability to pinpoint prey in the dark. But, for the harrier, hearing and vision work off each other and a movement through the reeds can first be detected through the ears before it is homed in on with the eyes.

The marsh harrier is as much a generalist predator as the red kite and sea eagle. Almost anything is taken where available: insects, crustaceans, fish, snakes, birds, small mammals ... In the British Isles the largest prey is an adult duck. Unlike the other harriers, the marsh harrier will also occasionally feed on carrion, its larger size and larger bill enabling it to compete amongst the other carcass-squabblers, the corvids, kites and buzzards. The marsh harrier specialises only when there is something to specialise on, when there is a glut in a particular prey species. But it is not reliant on a specific prey as other raptors – the vole specialists – are. Young moorhens, coots and rabbits are taken in the summer months. Also, partridges, skylarks, rats, small coypus, where available. In Kent, the marsh harrier is known as the *coot-teaser* (other local, archaic names include: *bald buzzard*, *white-headed harpy*, *duck hawk*, *moor buzzard*). It will repeatedly dive at coots and water-rail to exhaust and drown them, much as the sea eagle does. Marsh harriers have been known to barge the smaller Montagu's harrier off its prey. They have even been observed trying to knock an adult female pheasant off her feet to get at her chicks sheltering beneath.

They are more methodical too, marsh harriers, in their hunting than the hen and Montagu's harrier. Marsh harriers do not range as far when hunting (certainly nothing like as far as the long-distance foraging of the Montagu's harrier). The marsh harrier will work a patch of reed bed over and over, scrupulously checking it for the slightest movement, the slightest leaf shiver.

They rarely give chase. Marsh harriers rely on surprise and ambush, coming across prey unawares in the reed beds and ditches. But just as they hunt through the interplay of sight and sound, marsh harriers also hunt through a blend of search and flush, alternating their speed and height, leaning, like the other harriers, into a head wind to slow themselves, rising to take a closer look inside the reed bed, or, rushing low over a drainage dyke, hoping to surprise prey feeding on the marsh below the dyke. The male marsh harrier, his pale underside merging his outline with the sky, hunts more than the darker female over open spaces. The larger female is the reed-bed specialist. But both sexes prefer a broken landscape where they cannot be seen coming, where they can utilise landscape features – ditches, hedges, field edges – to surprise their prey. Often the harrier spots prey having already flown over it, performing a split-second somersault, tracking back to drop with wings held back and legs stretched out in front to make the strike.

Mostly it misses: most birds of prey miss most of the time. They are not the super-efficient predators we take them for. Five to 10 per cent is around the average success rate for raptors. Estimates for the marsh harrier are slightly higher, 5–17 per cent. One study in East Anglia gave a success rate of 27 per cent, though killing is hard to monitor because so much of it goes on out of sight in the deep reed beds. Young raptors chase everything and miss almost everything. They learn quickly what is suitable and what should be left alone. They learn to read the signals: coots flick their white tails at marsh harriers when they are still a long way off, saying, *We have seen you, we can see you coming, don't bother yourselves with us.* Skylarks, singing vociferously over the moor, are saying to the merlin, *Listen, I am loud and fit and can climb as far as the clouds if you try to pursue me. Save your energy, don't bother with me.* The myth is that killing is easy, that all the osprey has to do is turn up and fish will float to the surface, belly-up in submission. Most raptor prey is quick to adapt to predation. Raptors tend to hunt routinely, predictably. A marsh harrier will cover the same ground daily, often for years on end. Their prey know they are there: many small mammals seek out the cover of denser vegetation or make use of heavy rain (which grounds birds of prey) to do their foraging.

The marsh harrier kills with its talons, a seizure in the foot. The rear and front talons grip then squeeze, puncturing the prey from either side with the harrier's long sharp claws. If this does not result in death, prey can be stabbed with a talon or, more rarely, with the bill. Larger mammals are then skinned, the skin pulled down neatly like a sock from the head down over the rear legs. Avian prey is plucked then stripped of its flesh by the harrier's strong bill.

Enemies: heavy rain, bitterns, foxes, egg collectors, wild boar, mink ... Bitterns (who will eat anything, including marsh harrier chicks) are the marsh harrier's sworn enemy. As the raven is to the red kite, so is the bittern to the marsh harrier. Bittern-baiters, harriers will persistently mob any bittern that comes too close to their nest. The marsh harrier will swoop down at the bittern, veering up at the last moment as the bittern jabs its long spearlike beak at the harrier to ward off the attack.

I stand beside the caravan for a while, enjoying its shelter, hoping the marsh harrier will come back. But he has moved on, hunting the fields way over to the west, so I turn north and walk along a frozen farm track. I pass some outbuildings on the edge of a farm. I can see just an arm's depth into their open doorways, bridles on a nail hook, a lasso of blue bailer twine. Manure heaps beside the track are bulbous under the snow. I can still hear the sea. Between the track and the sea there is a narrow field with a thin windbreak of stunted birch trees.

I don't know why I stop there on the track, gazing through the swept-back trees at the muddy sea. I think I am just tired after walking into all that wind and I like the shape the dungheaps make under the snow, glandular, like giant puffballs. But then, out of the trees, racing towards me: a female sparrowhawk, the last bird I expect to see out here in the marshes. A burst of speed, then my binoculars find her and I can see the black and white patterning of her breast. It makes no difference to the hawk that I am here. What happens next carries on around me and close to me as if I'm not here at all. I stand on the track and watch the hawk fling itself at a blackbird, a glossy black male. The blackbird is on the ground, close to the fence that borders the track and the field. He is scolding hysterically, frantic with alarm. With its first stoop the sparrowhawk misses, lands on the ground beside the blackbird and proceeds to pounce at it, stabbing at the blackbird with its long yellow legs. The blackbird easily dodges the stabs. So the hawk tries again, lifts up, hovers a couple of feet from the ground, and flings itself once more at the blackbird. Again the blackbird easily steps out of reach and it seems to hold the advantage, the hawk has lost the element of surprise and it has too little space to use its speed. The blackbird, it seems, has no intention of taking flight. But the hawk will not give up, it keeps lunging at the blackbird until the blackbird has pressed itself right up beside the bottom of the wire-mesh fence, as if it is trying to squeeze itself under the fence. The hawk cannot risk flinging itself at the blackbird in case it misses and crashes against the fence. So it is finished: the hawk lifts off the grass and flies straight back into the trees. The blackbird, still scolding, shoots off low across the track and into the adjoining hedge.

I thought that was it and the sparrowhawk had gone. But she is suddenly flying back out of the trees and flinging herself at a large mistle thrush crossing the field between the track and the wood. The gap between the two attacks is barely a minute. The hawk seems exhausted, grabbing at the thrush, stabbing at it with its legs. Always missing, the thrush easily dodging the hawk's talons, feinting, checking, sidestepping away from the hawk. And once the thrush has twisted out of the momentum of the hawk's attack, it is over and the sparrowhawk turns away and flies back into the bare wood. For a few minutes afterwards I watch her burning in her hunger-rage through the trees, unsettled, skidding from branch to branch. Above the spinney, flocks of woodpigeon, like a weather system, stream over the trees on their way to roost.

Of all the British birds of prey the marsh harrier inhabits the most specialised habitat. It is a bird of the reed and the wet, the marshlands and fenlands, the border zone between land and water. The reed bed is the marsh harrier's backyard; in French it is *le busard des roseaux* (the harrier of the reeds). To hunt across such a landscape it needs to be able to listen deeply, see deeply and stretch deeply. So it has evolved the longest legs and largest feet of all the harriers and its greater size enables it to kill much larger prey than the hen and Montagu's harrier can.

Where it nests, in the deep reed beds, amongst bulrush, bur-reed, reed buntings and bitterns, is as remote and inaccessible as the peregrine or golden eagle's cliffside eyrie. A reed bed is the most difficult, unnavigable landscape to try and move through. I tried it once, late one summer (after the nesting season was over), walking through a huge reed bed on the Tay estuary. It was muggy and the river smelt of warm mud and rotting vegetation. On the path beside the river I almost stepped on a rabbit with foamy myxomatosis eyes. The rabbit sensed me and leapt awkwardly to the side, spraying dew from the long grass where it landed. I walked down a steep bank through thigh-deep thistles. Reed buntings were clicking all around me and through binoculars I could see the streaked patterns on their pale breasts and the distinctive black heads of the males. A female marsh harrier was cruising over the beds below me, her brown back the same colour as the reeds' seed heads. I could just make out her pale silver throat and nape. She circled briefly over a cornfield above the river, stirring up house martins that looped around her in agitation. Then she was out over the beds again, sweeping their expanse. In the distance, I watched her suddenly check her flight, turning right around. She swooped down towards the reeds, paused, then dropped into the reeds and out of sight.

I found a narrow path somebody had cut into the reeds and followed it. More like a tunnel than a path, the reeds completely dwarfed me and leant over to touch each other to form a roof. But then the path just stopped and I was confronted by a 10-foot-high wall of reeds in every direction. I tried to wade into them but it was impossible, I could not see where I was going and I needed a machete or stilts, or both. It was the most claustrophobic space I have been in, the reeds were so ungiving. So I gave up and waded back to the small clearing where the path ended. When I had stopped and caught my breath it was quiet inside the reeds. The place had its own acoustic reach, insects – the whirr of mosquitoes – were amplified inside the reeds. The floor of the bed was an oily black mud, I would have sunk into it if it wasn't for the broken reeds that lined the path and supported my weight.

Later I found a section of the path where the reeds were not quite so high and I could stand looking out across acres of reed bed. I enjoyed this new perspective, eye level with the tops of the reeds, at the same height as the marsh harrier cruising over them. I often seek out these sight-angles, try to get myself, if possible, at eye level with the birds. I had done this with the peregrines at the top of the cathedral tower in Coventry, the red kites on the hill above Llandovery and the golden eagles on the mountain tops of the Outer Hebrides. With harriers there is much less of a climb, you just need to find somewhere suitable to hide. Then wait, keep watch: the harrier's flight level is often as low as your eyes.

Perhaps more so than any other bird of prey the marsh harrier just needs to be left alone. They need the isolation – the strange quarantine isolation – of the reeds. They are the most fragile and flighty of raptors, easily spooked, easily disturbed. It does not take much for them to desert their nests: children splashing in a nearby pond, pleasure-boaters, a clumsy wildlife photographer ... The Danish ornithologist, Henning Weis, who studied both marsh and Montagu's harriers in West Jutland between 1913 and 1918, put it best when he described the marsh harrier as *a creature of infinite caution and wariness towards everything unknown*. One incident Weis records in his beautiful book *Life of the Harrier in Denmark* perfectly illustrates this caution in the bird. Weis was attempting to photograph a family of marsh harriers when the adult male bird spotted him crawling out of his hide in the reeds close to the harriers' nest. Weis at once destroyed his shelter but neither of the harrier pair landed in the nest again. Instead they dropped food down to the young as they flew over the nest. Eventually the adult birds persuaded the young harriers to abandon the nest site and move to a new location some distance away.

Henning Weis felt that this innate wariness was probably the only reason the marsh harrier still existed in Denmark, though he was pessimistic about its chances of surviving as a breeding species after the brief respite of the First World War. Once wildfowlers returned to the marshes again after the war was over Weis was sure the marsh harrier would be done for in Denmark. In Britain extermination of the species was more advanced. The last known breeding pair of marsh harriers were trapped in Norfolk in 1899. They had once been so common in that county they could have been named the *Norfolk Harrier*. And across the border in what is now Cambridgeshire, Whittlesea Mere, before it was drained, had been a mecca for the harriers, a place where oologists returned again and again to collect their eggs.

Gone by 1899; crept back in 1911; bred again in Norfolk in 1915. But the marsh harrier's status was so tenuous, so precarious, for many decades it was easily Britain's rarest bird of prey, rarer even (and you would not think this possible) than the red kite still clinging on in Wales. Numbers were boosted a little by wartime coastal flooding where land was deliberately flooded to deter invasion. Even more land was abandoned to the reeds along the Norfolk and Suffolk coasts after the devastating tidal surge of 1953, and by 1958 there were fifteen nests in Britain. Then came the pesticide years: eggshell thinning, convulsions, death by poisoning. By 1971 their numbers had plummeted to just one nest in the entire country, and the marsh harrier was the country's rarest breeding/not-breeding bird of prey again. Just as the red kite was exiled, banished to the remote Welsh hills for the first half of the twentieth century, so the marsh harrier, during the same period, was kept in check, contained within a small corner of the Norfolk Broads.

For decades the marsh harrier existed as a fragment of itself, a sprinkling of birds holed up – held up – in the marshes of Hickling and Horsey. The birds were held there in a quarantine of sorts, any attempt to colonise neighbouring areas being met with persecution, exclusion.

Marsh harriers were not resettled – not reintroduced – in the way that the red kite and sea eagle have been. They just came back of their own accord once the persecution abated. There are now over three hundred breeding pairs in Britain and they can be found in wetland habitats between the east coast of Scotland and the south coast of Dorset. Some of these, like the Montagu's harrier, are migratory, but recently, as winters have become milder, British marsh harriers have also begun to overwinter here.

In the end all landscapes tell the same stories, everywhere is layered with the same strata of clearances, displacements, resettlements. Around the Isle of Sheppey, in the mudbanks and creeks of the Medway, there is everywhere the legacy of enforced detentions, segregations and quarantines. Lazarets and prison ships: vessels on route to London from plague-infected ports were placed in quarantine in the creeks off Sheppey; hulks stuffed to the brim with French prisoners from the Napoleonic wars were also moored in the Medway. The prisoners died from smallpox, typhoid and cholera in their thousands and were buried in the surrounding marshes. Deadman's Island, lying just off Sheppey's west coast, became an island tomb. If you go to Deadman's Island today it is a place strewn with human bones, spilt from their graves by the sea's gnawing.

After the sparrowhawk encounter I head west across the fields towards a long, sinuous reed bed. The light is going and in the middle of one field I disturb a dozen swans, camouflaged against the snow. The swans rise heavily into the dusk. A merlin is here too, still charged with energy, still flickering intensely. He darts past me, skimming the ground, checking nothing is loose.

I wait beside the reed bed and watch as marsh harriers come in off the white fields making for their roost. Dark and slow, low-flying shapes folding into reeds. A large dark brown female with a reddish tint to her tail comes very close. At the last moment, when she sees me, she banks away, tipping herself over the edge of the reeds. Her pale cream face a faint light inside the dusk.

XII

Honey Buzzard

The New Forest

I reach the New Forest in the middle of August when the place is locked in its own microclimate of humidity. In deep bracken, beneath holly trees, I clear a space to pitch my tent. Bracken, cut and flattened, softens the floor to sleep on; in the Forest, for centuries, it was harvested as litter and bedding for animals. In the autumn, huge bracken stacks, tall as houses, grew up next to farm steadings as if the buildings had sprouted – bloomed – sudden strange appendages. Some of the stacks were so vast it looked as if an old rusty sun had been rolled into the farmyard and left there to burn itself out. And the smell! The great bracken-rich smell of the forest come out of the forest to linger in people's homes and hair and skin so that the smell pervaded everything.

There are holly trees all around my tent. Birch and oak are here too but holly dominates and I am glad of its shade and the way its thick screen hides my tent. Holly is hospitable like that, protective, nurturing: the humus it lays down helps oak and beech to get a footing in the degraded acidic soils of the Forest. A 'holm' in the New Forest is the name for a dense stand of holly. You see the holly holms rolling over the higher ground above the valleys like storm clouds snagged on the heath. I like that a holm can also, for a while, be a *home* to other species, to oak and beech and yew and rowan, nurturing them, protecting these trees from the voracious grazing of the Forest's deer and ponies. In places around my tent the ground is ankle-deep with brittle holly leaves, the floor crackles when I step on them. In among the leaves are dried whorls of pony dung, baked hard and white by the sun. When I flick the droppings with a stick to clear a space for my tent, underneath the baked exterior the dung is black and moist and pitted with tiny insect burrowing.

Holm is a common place-name in the Forest. As is 'hat', an older word for holm, and a lovely term which describes the prominent shape and stance of the holly stands sitting up on the high ground. Though such is the age of the New Forest, many of the 'holm' or 'hat' place-names no longer have hollies (or sometimes any trees) growing there, the original hollies having died and been replaced by open heath or other species of tree.

Often a place forgets its given name. Landscape is always changing and it can quickly become remote and unreconciled from the meaning of its name. So a holly holm becomes a home for something else, and 'hawk hill' or *Cnoc na h-Iolaire* (the eagle's hill) may not have seen a bird of prey for centuries. But the name remains to document the absence. And sometimes all you see of a place is what is missing.

The experience of absence was the most important experience of my journey. More often than not, that is the experience of searching for raptors: the birds are not where they might be, or not where you want them to be. Many birds of prey lead evanescent, hidden lives and many species are instinctively wary of man. Even buzzards, which are now quite common throughout much of the country, never let you get that close, always exiting their perch just ahead of your approach. Most of my time out looking for the birds was spent not seeing them, not finding them. Sometimes, as with goshawks in the Border forests or sea eagles in the Morvern peninsula, I went for days without seeing the birds. So I was confronted all the time by their absence and I came, gradually, sometimes reluctantly, to appreciate the experience of that absence as being crucial to experiencing the birds, to appreciating their rarity and fragility, to respecting their space, to acknowledging the distance the birds needed to put between us.

Everywhere I went – from Morvern's empty glens to the abandoned one-night houses of Wales – I witnessed absence, or, more accurately, the legacies of absence. A journey through Britain's landscapes is a journey through narratives of absence. The land is so much emptier than I had imagined. Sometimes I found it difficult to separate the two strands of my journey: the stories of the birds, and the stories of the landscapes I went to search for the birds in, were always interacting, always working off each other; human and raptor fusing, each one inhabiting the other. You cannot separate the story of Britain's birds of prey from the birds' relationship with man. That relationship *is* the birds' story. So the narratives of absence I kept coming across in the wider landscape often, unavoidably, tangled themselves with the stories of persecution, removal and extinction that mark the narrative of so many of our birds of prey.

I went to the New Forest to try to get better at not seeing, not finding the birds. I wanted to be more patient, to not always be impatient to try to tick the birds off with a sighting. I wanted not to worry too much if I did or didn't find the birds. I wanted to appreciate more the experience of not seeing them. Historically, traditionally, honey buzzards have bred in the New Forest. Perhaps they were there when I visited, perhaps not; I was determined to not really mind, to let go of minding. Some birds live cryptozoic (hidden) lives and I wanted to respect that. My aim was simply to immerse myself in the honey buzzard's forest habitat and spend some time amid the possibility of the birds.

I am drawn to cryptozoology, the study of hidden animals, the searching for creatures that may not (and are unlikely to) exist. Most scientists think cryptozoology a nonsense and a waste of time, searching for Yetis, Loch Ness Monsters and the like. But I'm drawn to cryptozoology as a depository of metaphors, that searching for something that may not be there, the questing after absence. Cryptids (the creatures that cryptozoologists study) are things of absence or such elusiveness that they only exist beyond our reach or ken. Sometimes on my journey it felt that I was searching for birds that had slipped beyond my reach. Birds of prey can be so elusive, so unattainable, they can feel, at times, like cryptids themselves.

If I had to pick one bird of prey to represent – to epitomise – absence, it would be hard to choose from the roll-call of candidates. So many of our raptors are synonymous with absence. Several, still, are conspicuous by their absence. But if I had to choose, I would choose the one that lives the most cryptozoic life. The honey buzzard is a bird of such elusiveness and strangeness that it teeters on the brink of myth, a bird whose presence here is so short-lived and secretive it is barely here at all.

* * *

The first visitor to my new home – to my clearing in the bracken – was a robin. I heard its tiny feet on the leaf litter and noticed the bracken fronds shifting as the bird brushed against them. The robin hopped closer, paused, tilted his head towards me then skittered across the leaf fall, his feet over the dry leaves making a soft scratching sound.

After the robin had gone there was a long stint of quiet. I sat outside the tent until it was dark. Then the owls began. Right over my head the first tawny owl sent out a sharp, piercing *k-wick*, answered by a long-drawn-out *hal-loo*. For the next half an hour the owls took over and the noise of their calling was all around me. The tent caught and amplified every sound that touched it and I heard one owl leave its perch and zip over the top of the tent to land in the nearest holly tree. There must have been only an inch or two between my face and the owl as it skimmed over the canvas. Then it was quiet and I slept a little and woke again at 2 a.m., woken not by any noise but by the silence. It was so still, so strangely quiet, as if the forest was listening to itself.

At dawn, all new sounds: a blackbird scuffle, a jay's scratch, a woodpecker's yaffle. I get up and wash my face in a rusty brook, stirring up clouds of iron breath. After washing, I push, by mistake, through a cove of spiders' webs strung between pine branches, feel the threads prickling, sticking to my damp hair and skin. Beside the brook are several birch trees, their trunks so wrinkled with age they look like oaks.

The rest of the day I spend wandering through the Forest. For long periods I just sat, waited and listened. And wonderful things came to me that way: a nuthatch, a lesser spotted woodpecker, a squirrel-shower (beech leaves spluttering down on top of me from where a squirrel leapt). Tree stumps were dinner plates for pine-cone seeds left there by the squirrels. The air among the spacious rows of pine was cooler where the breeze had more room to flex. The pink skin of Scots pine showed through the cracks in its fissured bark. Ponies stood in swishing pools of shade amongst the trees.

The New Forest is such a vulnerable, fragile place, encircled – periodically threatened – by development. It is an environment which endures massive pressures on it, from recreation to livestock grazing; in places the forest is so sparse, grazed down to its bare knuckles by the ponies and deer. Only the unpalatable species remain: wood spurge, wood sorrel, butcher's broom ... The forest understory is such a depleted space, you can see a long way – unnaturally far – through the trees. What is abundant – what is luxuriant – in the Forest is the great diversity of bryophytes, the mosses which love to fur the moist south-western side of trees and the lichens which blotch and crust the rest. And everywhere too in the Forest there is an abundance of decaying wood. Old trees take so long to die and in their long death the Forest's ancient trees are home to many insects. All the dead-wood tenants are here: flies, beetles, bees and wasps. And they are here like nowhere else in the country, the diversity of invertebrate species is quite exceptional. Thirty species of bees and wasps alone in the New Forest and, of the wasps, the common wasp is found here, also the German wasp, the red wasp, the tree wasp and Norwegian wasp.

Wasps: they are the reason the honey buzzard is here. They are the reason the bird migrates in summer from sub-Saharan Africa into Europe and Russia. Wasps are the reason too why the honey buzzard lives such a cryptozoic, secretive life and why it is so unlike any other bird of prey. An old name for the honey buzzard was *bee-hawk* and that is a more true description of what the honey buzzard is, though *wasp-hawk* would be more accurate still. For, despite its beautiful name, the honey buzzard does not eat honey, nor is it a buzzard. What it does eat – what it loves to eat more than anything – is wasp larvae.

Unlike any other bird of prey is the way its claws are blunt and almost straight (not sharp and curved like other raptors). For the honey buzzard is a walker, a burrower, a digger-out of wasp nests from underground, a bird-badger. And it is a good walker, not awkward on its feet like other large birds of prey. MacGillivray noticed from his dissecting table that the bee-hawk's claws were *long, rather slender, arcuate, less curved than in any other British genus*. He noticed too in the specimen of a young male honey buzzard killed near Stirling in June 1838 which came into his hands on the 9th of that month, *when it was perfectly fresh* that its *soles were crusted with mud or earth; the claws very slightly blunted*.

Also unlike other birds of prey in the way its bill is more delicate. It needs to be, to enable the honey buzzard to carefully extract the wasp grubs from their cells. For the same reason, the bird's tongue is also highly distinctive, fat and tubular, designed for prising – perhaps sucking – the larvae from their chambers. The adult honey buzzard feeds its young like this, plucking out wasp grubs one at a time from the wasp comb then presenting the grubs to its chicks. It is a delicate, tidy procedure, the comb is handled carefully until every grub has been removed. Only then is the comb discarded, often trodden down into the detritus of the nest.

Even the honey buzzard's internal organs (MacGillivray would love this observation) are distinct from other birds of prey. Their gizzards are lighter and their small intestine much shorter than in most other raptors because the soft wasp larvae are more easily digested than the flesh and bones and fur that many other birds of prey consume. Also, wasps are the reason the honey buzzard rarely regurgitates pellets. It rarely needs to, because, again, the softness of the insect larvae is easy to digest, unlike the indigestible matter – the bones and feathers – which other birds of prey expel in their pellets.

Wasps are crucial, they are integral to what the honey buzzard is. But wasp larvae are not the only thing honey buzzards eat, the larva is not always available. In Britain, when the first honey buzzards arrive from their migration around the middle of May, wasps are largely inactive. So the birds top up their diet and, importantly, the females increase fat reserves before laying, by feeding on nestling birds (especially pigeon squabs), frogs, lizards and other insects. Dungheaps are favoured resources for excavating grubs and worms. Beetles, weevils, earwigs, ants … have all been found in a honey buzzard's stomach. The stomach of the male specimen that MacGillivray dissected *was filled with fragments of bees and numerous larvae, among which no honey or wax was found.* Bee nests are frequently raided by honey buzzards. In the British uplands, where bumblebees are common on the heather moors, the bees are a significant prey species for the birds. Sometimes young honey buzzards become gluey with honey that leaks from the combs brought into the nest by the adult birds, leaves and twigs stick to the legs of the chicks as they wander about the nest. In poor wasp years, bees can provide an important supplement. Also frogs, which are usually skinned first by the adult birds before they are fed to the young.

But even when wasps are inactive, the honey buzzard is thinking about wasps. It has to think about them – to anticipate the wasps – in order to survive and rear its young. So after laying and during the long period of incubation the male honey buzzard becomes a map-maker, a cartographer of wasp nests. He does not raid the nests during this period as they are not sufficiently developed. Instead he makes a survey of his patch, recording – storing – the locations of all the wasp nests he can find. Later in the season, he will draw on this memory-map to return and plunder the nests when they are more advanced and stocked with fattening grubs.

But all birds of prey are map-makers. They rely on intimate knowledge of landscape to hunt, routinely returning to the places where they know prey can be found at certain times of day. Routine is everything. Landscape is memorised, landscape *is* memory. But the honey buzzard's map is more enhanced in scale. It deals in the minutiae of place, in the minuscule world of invertebrates. The woodland clearing, the forest ride and forest purlieu, these are the honey buzzard's theatres, its zones of interest. Dense forestry plantations are no good, they suffocate everything. The honey buzzard needs woodland that is light and roomy. It needs glades and tracks, woods that intersperse themselves with openings. The honey buzzard sits in a tree on the edge of these clearings, keeping watch, static hunting. And when it spots a worker wasp heading back to its colony, the honey buzzard slips from its lookout branch like a shadow unhooking itself and follows in the wasp's wake, tracking the wasp back to its nest.

But if the weather closes in, smudging everything, then wasps can be hard to track like this. And it's then that the honey buzzard's memory-map of wasp-nest locations becomes a lifeline. By storing the locations of wasp nests early in the season, the honey buzzard is laying down a cache of knowledge. And this enables the birds to be more climatically resilient than it's often assumed they are. It enables honey buzzards, for instance, to breed just as successfully in the British uplands (where it is generally cooler and wetter) as they do in the more benign lowlands. If wasp numbers are seriously depressed this can impact honey buzzard breeding success. But wasps are also more climatically resilient than we assume and the insects have been observed to forage even in heavy rain.

Perhaps more than climate – more than temperature and precipitation – soil consistency is what limits the distribution of honey buzzards. The earth needs to be diggable, friable. The honey buzzard is a miner of wasp nests, it needs to be able to extract the nests from the ground. So heavy clay is no good, hard, arid ground is no good for honey buzzards. Even where wasp numbers are high, as in Mediterranean countries, honey buzzards are scarce because of the difficulty of getting at the wasps in such a dry, baked landscape. Light sandy soils are good, ground covered in a thick mulch of pine needles is good. The bird's digging instinct is so strong that young honey buzzards are known to scrape holes in the bottom of their nest, occasionally to such an extent that their digging undermines the stability of its structure.

Nothing like a raptor, it digs like a dog, scratches away, deeper and deeper until the bird can sometimes disappear into its own hole. So absorbed in the task of excavating the wasps' nest, it is said that you can walk right up to a digging honey buzzard. Though this is not advisable: where a honey buzzard digs there are likely to be furious wasps, made more furious by the fact they cannot do anything about the honey buzzard and have been known instead to turn their fury on anything else to hand, ponies, dogs, passing ornithologists …

The honey buzzard seems to work at its diggings with impunity. Wasps will swarm about the bird but appear not to harm or deter it greatly. It's possible the bird releases a chemical to calm the insects. The thick scale-like feathers on its face (which other raptors do not possess) may also offer protection from stings (like trying to sting through a pineapple's skin). Even so, sometimes a honey buzzard will be driven back by a ferocious onslaught from the wasps. But after a respite of head shaking, shrugging and intensive preening the honey buzzard usually resumes its pillage. Sometimes wasps and bees are snapped at by the bird, plucked from the air (often decapitated) and eaten. The whole process of digging out the wasps' nest can last for hours, with the nests often awkward to get at, lodged under layers of grass and tree roots. And once the nest is accessible the process of feeding from it can last for days, with the honey buzzard returning again and again to retrieve chunks of grub-rich comb.

Nothing like a raptor, but not exempt from the raptor's fate. You would think a chiefly insectivorous bird (a predator of wasps no less) might be immune from persecution, even welcomed. But when the war against birds of prey was at its height during the nineteenth and early twentieth centuries, there was little discrimination between raptor species. The *Birds of Hampshire* notes that twenty-four honey-buzzard nests were recorded in the New Forest between 1856 and 1872; at least twenty of these were robbed of eggs or young birds and the adult honey buzzards killed.

* * *

One evening in the New Forest, walking back to my tent through the trees, I heard a bird of prey calling. It was a loud, piercing call, agitated, persistent, a long whistling note, rising in pitch. It confused me, I couldn't identify the caller. It was roughly the same volume and pitch as a common buzzard but it sounded different, deeper, more plaintive. At least, it sounded different from the soundtrack of common buzzards I was so used to hearing at home. I walked towards the call, trying to get a bearing on where it was coming from. I searched for half an hour, pausing and listening, scanning the branches of tree after tree. Nothing. I found it impossible to pinpoint the sound in the deep foliage of the trees. Instead I tried to memorise the call (*ke-yeeeep ke-yeeeep …*) and when I got home I spent a whole morning of confusion, listening to recordings of common buzzards and honey buzzards, trying to convince myself that what I had heard was a honey buzzard.

I went to the New Forest determined to be more patient about not finding the birds, but it would be dishonest to say that I didn't mind not seeing a honey buzzard. I longed to see one. And I explored as much of the Forest as I could in my search for the birds, even following wasps to see if they would lead me to their nest (they never did). I tried, but I could not shake that longing to find the birds. Not seeing them only made me dream of honey buzzards incessantly. Perhaps that is what a cryptozoologist is, somebody who dreams about the same creature over and over again; perhaps that is what a cryptid is, a creature that hides inside dreams.

After I got home from the Forest I read a fascinating paper hypothesising that the juvenile plumage of honey buzzards has evolved to resemble the plumage of common buzzards. Goshawks predate honey buzzards (especially when pigeons and rabbits are scarce), but goshawks are more reluctant to predate the sharper-clawed, more aggressive common buzzard and so juvenile honey buzzards, to deflect the risk of predation from goshawks, have, through a process of Batesian mimicry, adapted their plumage to more closely resemble their more robust cousins. The two species – common and honey buzzard – are difficult enough to tell apart; that the honey buzzard should try to absent itself even more by hiding in another bird seems somehow apt for this most cryptozoic bird of prey.

But how absent are they? Perhaps not as much as we tend to think given the honey buzzard is so good at being overlooked. MacGillivray wrote in his book *The Natural History of Deeside*:

> Even on the border of the most frequented paths are many things travellers have passed by unheeded or unexamined.

The honey buzzard, despite its size, is an easy bird to pass by unnoticed. Still, it is only here in fairly small numbers, roughly sixty-plus pairs. But there is room – plenty of room – in these islands for many more. Anywhere that is diggable, friable and waspy is good for the wasp-hawk.

* * *

The sections I keep returning to in MacGillivray's journal of his walk to London are the lists of flora he recorded along the way, each night writing up the Latin names of all the plants he'd seen that day. Between the Bridge of Dee and Upper Banchory, alone, he notes down fifty-seven species of plant in flower and sixty-two out of flower (all of these committed to memory as he walked along).

There are moments in his journal when he seems to conjure beauty out of nothing, out of nowhere. For instance, a quarter of a mile on the road out of Buxton he pauses to listen in the clear morning air to larks singing and out of the corner of his eye he sees a plant he does not recognise, a musk thistle (*Carduus nutans*), *the most beautiful thistle*, he writes in his journal, *which I have yet seen*. Soon after this he finds the common carline thistle and, at that point, *everything*, he writes, *conspired to render me cheerful*.

Some of the plants MacGillivray collected on his walk to London have survived and are held at the University of Aberdeen's Herbarium. I spent a whole morning in the Herbarium going through the collection, looking at the sheets of plants exactly as MacGillivray had left and labelled them almost 200 years ago. Hundreds of specimens from a lifetime's botanical wanderings and all of them presented and organised with such clarity. I found plants he picked on his 1819 walk to London and many he collected from the high Cairngorms in the summer of 1850, just two years before his death. Several of the mountain ferns and grasses still had residues of peat tangled in their roots. Some of the flowering plants, remarkably, still retained the colour in their petals. The purple flowers of *Statice armeria* were paled and rusted but still had hints of pink in them. The note that MacGillivray wrote beneath each specimen (his handwriting straight and neat across the page, no archaic floridness to it) was usually the plant's Latin name followed by the place and date he had collected it. I thought that if I spent enough time with his collection I could draw a map of all the places he went in search of plants. I could trace MacGillivray's movements across the land, across the years, that way. That would be how I would write a biography of William MacGillivray, his life told through his plant cuttings, his botanical wanderings …

– Deschampsia Flexuosa: Gathered on Ben Nevis, Inverness-shire 16th September 1819, by Mr MacGillivray.
– Statice Armeria: Summit of Benvrotan near one of the Sources of the Dee, 10th September 1819.
– Statice Armeria: By the Dee, 8 miles from its mouth, May 1819.
– Asplenium Adiantum nigrum: Wall of the Marquis of Abercorn's orchard at Duddingston, near Edinburgh, 1st January 1820, 1 A.M. moonlight.
– Lycopodium Selago: Lochnagar, 8th August, 1850.
– Carex Saxatilis: Corry of Lochan uain, Cairntoul, 12th August, 1850.

Dear William, I am sorry for their desecration. After twenty minutes of searching the New Calton cemetery in Edinburgh I have found William MacGillivray's grave. For a while I was concerned I might not find it at all, several of the headstones in the churchyard have keeled over and I was worried his might be one of these. But I found it in the end, a great pink-flecked slab of granite with an Iona cross set in a Celtic scroll at its head. The long inscription a little faded inside the granite, though still legible:

> In memory of William MacGillivray, M.A., LL.D., born 1796, died 1852. Author of A History of British Birds and other standard works in Natural Science; Professor of Natural History and Lecturer on Botany in Marischal College and University from 1841 to 1852. Erected in 1900, together with a memorial brass in Marischal College, Aberdeen, by his relatives and surviving students, who affectionately cherish his memory, and by others desirous of doing honour to his character as a man and to his eminence as a naturalist.

It is a fine spot, his grave, with views out across to Arthur's Seat, a bird-busy place with gulls coming and going from off the Firth of Forth. But the shock is what has been done to his gravestone. Somebody has hacked away at the granite and stolen the brass plaque that had been fixed to the base of the stone. I have seen photographs of it before it was vandalised, the plaque had been beautifully cast with the image of a golden eagle modelled from one of MacGillivray's own paintings of the bird.

I kneel down and run my fingers along the jagged edge of the granite where it has been hacked and chipped. It is still possible, just, to see the eagle's outline imprinted on the stone. It's as if the shape of the metal cast has branded – seared – the stone beneath. *Sear*: flip the verb into a noun and a 'sear' is also the 'foot of a bird of prey', from the Old French *serrer* to grasp, lock, hold fast ... I can see in the stone the line of the eagle's back, the curve of its chest, its long tail. All that is left is the faintest trace of the bird, a ghost eagle scratched into the stone, a hieroglyph.

XIII

Hobby

Dorset

Anything to make him feel less absent. With the brass plaque on his gravestone stolen, presumably melted down, I went in search of MacGillivray's painting of the golden eagle which the plaque had been modelled on. There are over 200 of MacGillivray's paintings held in the library of the Natural History Museum in London. Most of them are watercolours of birds and there are also several paintings of different species of fish and mammals. MacGillivray's intention was that the bird paintings would be used to illustrate his five-volume *History of British Birds*, though, tragically – and it is tragic because the paintings are quite exceptional – in the end the watercolours were left out of his books because MacGillivray could not afford the printing costs of including them. MacGillivray's son, Paul, donated his father's paintings to the Natural History Museum in 1892, where they sit in storage and seldom see the light of day.

But that is MacGillivray for you, a man eclipsed, the legacy of his work eclipsed by his contemporaries, Darwin, Audubon and Yarrell (William Yarrell, whose own more accessible, more popular, *A History of British Birds* was published in 1845). So that MacGillivray himself could be said to have become a type of cryptid, virtually unknown, hidden from view, forgotten. When I started to research this book I had no idea who William MacGillivray was, but then a brief quotation led me to his work on birds of prey, *Descriptions of the Rapacious Birds of Great Britain*. And of course the book was out of print and impossible to find but, when I finally tracked it down and began to read it, I could not stop reading. And soon I was trying to find and read everything I could by MacGillivray. His descriptions of the natural world felt chiselled from hours of careful observation. His account of watching hen harriers in his book *Descriptions of the Rapacious Birds* led me to his great work, *A History of British Birds*, where I sought him out and found him kneeling on the side of a hill, watching, this time, a different pair of hen harriers:

> Kneel down here, then, among the long broom and let us watch the pair that have just made their appearance on the shoulder of the hill … How beautifully they glide along, in their circling flight, with gentle flaps of their expanded wings, floating, as it were, in the air, their half-spread tails inclined from side to side, as they balance themselves, or alter their course! Now they are near enough to enable us to distinguish the male from the female. They seem to be hunting in concert, and their search is keen, for they fly at times so low as almost to touch the bushes, and never rise higher than thirty feet. The grey bird hovers, fixing himself in the air like the Kestrel; now he stoops, but recovers himself. A hare breaks from the cover, but they follow her not, though doubtless were they to spy her young one, it would not escape so well. The female now hovers for a few seconds, gradually sinks for a short space, ascends, turns a little to one side, closes her wings, and comes to the ground. She has secured her prey, for she remains concealed among the furze, while the male shoots away, flying at the height of three or four yards, sweeps along the hawthorn hedge, bounds over it to the other side, turns away to skim over the sedgy pool, where he hovers a short while. He now enters upon the grass field, when a Partridge springs off, and he pursues it, with a rapid gliding flight like that

> of the Sparrow Hawk; but they have turned to the right, and the wood conceals them from our view. In the meantime, the female has sprung up, and advances, keenly inspecting the ground, and so heedless of our presence that she passes within twenty yards of us. Away she speeds, and in passing the pool, again stoops, but recovers herself, and rising in a beautiful curve, bounds over the plantation, and is out of sight.

There is a warmth and intimacy to the writing, also a passionate energy; MacGillivray's voice, in this respect, feels different from many of his Victorian contemporaries'. He draws you in: *Kneel down here, then, among the long broom and let us watch …* He himself is so palpable, so present in his work; you cannot help but warm to him. MacGillivray's friend and correspondent the ornithologist James Harley wrote of MacGillivray's writing:

> Having for several years past paid considerable attention to the Ornithology of the British Islands, I would venture to recommend the beautifully written and elaborate History of British Birds, Indigenous and Migratory, by Mr MacGillivray, as being the best work in that department of science yet published. There is a peculiar mountain freshness about Mr MacGillivray's writings, combined with fidelity and truths in delineation, rarely possessed by Naturalists, and hitherto not surpassed. To the Ornithological Student this charming History of British Birds ought

> to become a hand-book – to the observer, a
> companion – and to the rambler in woods and wilds,
> a guide and pole-star.

MacGillivray wrote in the journal he kept of his walk to London a sort of manifesto of how he intended to write the journal. In these notes to himself, these memoranda, you glimpse a little of where that *peculiar mountain freshness* to his writing is drawn from:

> – I shall write it with as much freedom as if I were convinced that no person should ever read it. At the same time it must be so written that others may readily understand it …
> – I must avoid obscurity …
> – I despise opinion unsupported by reason, detest bigotry, and rejoice in persecution, that is in being persecuted, but not in exercising any authority, much less persecution, over others …
> – But while I write with the intention of benefitting others, and of gratifying my own vanity, I also write from the conviction that my notes will be useful to myself on many future occasions, yea even unto the day of my death.
> – Partly from the instigation of vanity, and partly from other motives, I refrain from laying down any general or particular plan either for my journey or journal. Only I shall drink a mouthful from the source of the Dee, and give three cheers to myself on the top of Ben Nevis; and till that time keep a

regular list of all the plants and birds which may occur.

Hobby, from the Old French *hober*: to move, to stir, to jump about. The hobby is the most kinetic bird of prey I know. It is all zip and dash and rushing speed, like a frantic whirligig of the heath. Its scientific Latin name, *Falco subbuteo*, simply means 'smaller than a buzzard'; its Greek name, *Hypotriorches*, translates as 'somewhat near a bird of prey' (the hobby's original scientific name was *Hypotriorchis subbuteo*). Both definitions are so wonderfully hazy in their description of the hobby, strangely, paradoxically, they actually capture the difficulty of grasping this small falcon as it rushes past you in its blur of speed.

A hop and a skip from the New Forest and I cross the border from Hampshire into Dorset. I head east from Wareham down a narrow road across the dusty heath to the Arne peninsula, a fist and outstretched thumb of land that juts out into Poole Harbour.

Walk due north from the tiny village of Arne, following the track across the heath, till you come to a scattering of birch and oak which lines the top lip of Arne Bay. Enter the trees and wade through the undergrowth of bracken, pausing when you reach the first lunar white birch. You will know it because of the way its trunk glows like a bar of light inside the wood. Turn left at the birch and you will come to a great oak draped in honeysuckle and ivy. Try to go there when the honeysuckle is in flower, when its flowers glow like small pale suns amongst the oak's dark leaves. You will know the oak for sure because one of its huge arms has broken off and is lying amongst the bracken beside the base of the trunk.

That is where, in a shallow hollow next to the huge oak, I slept that first rainy, windy night. A night of not much sleep, woken by rain and by deer sheltering from the rain; sika deer with their white mottled red-brown coats and their high-pitched alarm calls sounding like a bird's screech. I can't remember if I was asleep when the branch from the oak cracked then split and crashed to the ground, but it was a heart-thumping moment and I was up and out the tent within seconds peering into the gloom till I noticed that something was wrong with the tree, that it did not look right. Then I saw the branch sticking out of the bracken like a half-submerged wreck and the crashing noise suddenly made sense and, a little less spooked, I crept back inside the tent. Earlier, before the wind got up, I heard music coming from a bar across the harbour. Then, much closer, a churring noise as nightjars began to whirr across the heath.

MacGillivray's paintings have been arriving all afternoon in huge green boxes wheeled down on trolleys from the museum's storage. Wearing vinyl gloves, I lay the paintings out on a table one by one and write my notes in pencil.

- Here is his kestrel, a female, flaring red. He has the bird's bright chestnut colours, the detail of its down feathers, the cupped disc beneath the eye, the steep cliff-drop of its beak, the beak's black tip, the marble blue of the beak below the cere, the faint yellow of the cere itself, a black glint to the claws, a thread of ivy winding round the rock the bird is perched on; the kestrel poised, about to take off.
- Here is his ptarmigan with its feathered feet and semi-winter dress, painted 'from an individual purchased in the Edinburgh market October 1831'. What is beautiful about his ptarmigan is the way he captures those feathers dissipating as they ermine from gold and grey and black to northern white.
- Next, a storm petrel, 'from an individual caught off the Isle of May, 1832'.
- Then a pair of buoyant, bobbing wrens, 'shot in the Pentland Hills in the Summer of 1832'.
- His water ouzels, 'from individuals shot near St. Mary's Loch, September 1832', one in its first year of plumage and one young bird after moulting, the red-brown colours of its chest just starting to leak through.
- Sparrowhawks: an old orange-eyed female, long stretching talons, long legs, grey-black back.
- A tall, broad-chested, yellow-eyed female peregrine and a juvenile (I think) beside her with a reddish chest and heavy chainmail streaks across its breast.

- His merlins are precise: the blue of the male's back, the black wing and tail primaries, the orange blush of his chest; the female's lesser reds, her brown back, longer wings, longer tail; the juvenile's crenulated brown and chestnut patterning, mantling a ring ouzel which is almost the same size as the merlins.
- Ah, and here is his honey buzzard, that male 'taken near Stirling in the beginning of June 1838'. What a dark brown bird it is, so uniformly brown except for the grey above the cere and the light russet plumage above its legs.
- A red kite, another kestrel (the two red raptors next to one another).
- An osprey with the incomplete painting of the fish it's caught, just a pencil sketch of where the fish should be. One of the osprey's talons is also unfinished, lodged into the unfinished fish.
- And here is the painting of the golden eagle I was looking for. It is the last to emerge from storage. In fact, two paintings of eagles. One, like the osprey, is incomplete, only the cere and bill are painted. The other is of the eagle MacGillivray's gravestone plaque was modelled on, the eagle's thick talons embedded in a rabbit, the rabbit's right eye wide and bright. The painting feels a little artificial: the eagle is standing too tall, not mantling its prey, and the rabbit, despite the eagle's talon puncturing its neck, seems very much alive. A juvenile bird with white feathers showing on its tail and legs and

> wing, every feather delineated, the base of its tail
> all white except for the black trailing band.
> Demerara hues in the feathers along the eagle's
> crown and nape.

Hobby, *hober*: to move, to stir, to jump about ... I never learn; better to stay put and wait for the birds to come to me, don't wander aimlessly about the heath all day. Which, of course, is what I do, and do not mind because I see so much that way. I am looking for the flash of white of a hobby's throat lit up against the pine greens of the heath. And so everything that is white draws me in: egrets, an albino sika deer, black-headed gulls, a pair of spoonbills in the shallows, shelduck further out in the bay. In the woods I disturb a herd of deer and walk through the warmth of where they had been.

What am I looking for? A bolt of speed, a disturbance amongst the swifts. Something 'smaller than a buzzard', something 'almost like a bird of prey'. The hobby's jizz – its feel – the gist of it: somewhere between a peregrine and a swift. It has the peregrine's black moustachial stripe, the falcon's sharp-angled wings; it has the swift's sickle, boomerang shape in flight. Prey: anything from a flying ant to a turtle dove. But especially dragonflies, moths, beetles, small birds, willow-wrens, chiffchaffs, tits, pipits, hirundines (swallows and martins), swifts, bats ... The hobby is the only British bird of prey agile and fast enough to catch a swift, a bat, or a swallow in mid-air.

Just as the honey buzzard times its arrival and nesting to coincide with wasps, the hobby breeds in sync with the glut of young hirundines and July's crop of dragonflies. That the hobby (like the honey buzzard, osprey and Montagu's harrier) has to get out by autumn, migrating to sub-Saharan Africa, shows just how insectivorous a bird it is. Diet dictates migratory behaviour: the general rule of thumb is those raptors that feed on warm-blooded prey stay put, those that feed on cold-blooded species get out for winter as their prey dries up in the north. But why come back in spring? Why bother with such a long, hazardous migration twice a year? Why not stay in their wintering grounds, where it is generally warmer, where there is still food even for the insectivorous raptors in winter? Because the north in summer can be a land that is overbrimming, a place which offers much better opportunities for these birds to breed successfully. The pull of the north in spring is overwhelming. Hobbies have been recorded migrating northwards at an average speed of 71 kph (the return journey south in autumn is usually much slower than this). The journey – the awful journey – running the gauntlet of storms and Maltese gunmen who will shoot any migrating raptor within range, is not the point. It is getting back to the fickle north with its promise of all that light to work in.

Diet dictates migratory behaviour but also – for MacGillivray – diet dictates everything. To properly understand the bird, he felt, you needed to understand the mechanics of the bird:

> ... although I have selected the digestive organ as pre-eminently worthy of attention, I have not done so because I suppose them capable of affording a key to the natural system, but because the structure of the food determines not only the form and structure of the bird, but also the greater part of its daily occupations.

Wasps and other hymenoptera (bees especially) are integral to what the honey buzzard is; dragonflies and hirundines to what the hobby is. Especially dragonflies: in the British Isles, in recent decades, hobbies have been both increasing their numbers and spreading northwards out from their traditional heartland, the downlands and heathlands of southern England. Increase in dragonfly numbers has been an important factor in enabling hobbies to spread like this. There are roughly 2,800 hobby pairs now breeding here. Blink, and you will miss them.

The hobby in its winter quarters becomes a rain bird. It follows the rain, tracking thunderstorms and bands of rain that sweep south through southern Africa at this time of year. The hobby is much more insectivorous here than it is in its summer breeding territory. It can afford to be, can afford to hunt and eat less without the burden of young to raise. So the hobby tracks the rain, flies through the rain, flitting from rain cloud to rain cloud. Because the rain releases – induces – insects in their millions: locusts, flying ants, termites ... The hobby is just one among many species of raptor that follows these insect blooms. After the rains come grasshoppers, beetles, dragonflies, small birds such as queleas, pipits and cisticoles.

I spend a lot of time walking along the north shore of the Arne peninsula. A fringe of gorse and birch separates the shore from the heath beyond. I enjoy slipping through this curtain of trees and scrub, from heath to shore and back again. The heath feels as remote as anywhere I have been on this journey, yet I just need to step through a gap in the gorse, breach the border, and there are the office blocks, the flats and cranes of Poole, and a parting in the trees is suddenly filled by the vast bulk of a ferry coming into dock. The heath is like a pulse of wildness nudging the fringes of the town. A deer's alarm bark is followed by a lorry's warning alarm as it reverses along the quayside. I go from the blue-green glint of dragonflies flickering over the heath to sunlight glinting off the windows of houses in Poole.

MacGillivray loved this zone, where the city gave way to the countryside. Faced with a disappointment or difficulty his instinct, usually, was to walk to the edge of the city and dip himself in the surrounding fields. It's not that he was fleeing the city, he simply walked out to its margins, took stock, reorientated himself, then walked back into town again. Often he would jot down some notes on the geological features he observed at the edge of a city (on an excursion to the south of London he was delighted to have the opportunity to study the chalk there, a geological district he had never seen before). On a visit to Glasgow in 1833 he finds the museum he wishes to visit closed and writes in his journal:

> I then proceeded to the College, whence I was, however, obliged to return, the Museum not being open. So I had recourse to Nature, as I often have under more grievous disappointments, and betook myself to the margin of the city ...

Recently some birds of prey have come to live amongst us in our towns and cities in ways that they have not done for many centuries. Peregrine and red kite, most notably, can now be seen across many of our urban centres. Some of the most memorable encounters I've had with birds on this journey have been in built-up areas. I once spent a morning standing behind a huge oak doorway that led out onto the flat roof of a church tower in the centre of a large town in South West England. Three spy holes had been drilled through the door. Before I put my eye to one of the holes I could smell the falcons: it was a warm day in June and the place reeked. So this is what a raptor's eyrie smells like ... It was as if I had climbed to the top of a cliff, poked my nose over the brim of a nest and inhaled the pungent carcass stench of the place.

I put my eye to one of the holes in the door. The roof was littered with feathers, small fragments of bone; I could just make out a pigeon's leg. And there, in the far corner, were three young peregrines, well feathered, the odd clump of down sticking to them like candyfloss. The birds were only 6 feet away from me behind the door. They were lying flat with their chins pressed to the floor, long necks stretched out behind their heads. One of the birds was panting in the heat.

The adult female was perched above the roof on a corner spire. Dark bars drawn across her breast. Heavy, muscular chest. Bright yellow talons: scaly, reptilian. The black moustachial stripe running down her cheek was like a long drip of wax. She kept twisting her head back, glancing up at the sky.

What happened to Arne was another form of clearance. Just as a slice of the moor above Bolton was commandeered during the war as a decoy to draw bombing raids away from the ordnance factory at Chorley, they also built a decoy on the Arne peninsula to draw the Luftwaffe away from the crucial Royal Naval cordite factory at Holton Heath to the west of Poole. Arne was a small rural community: if the decoy worked the majority of bombs would fall, it was hoped, on heathland, oak woods, saltmarsh and that rare, beautiful space where oak woods hem a saltmarsh.

On the night of 3 June 1942 the Luftwaffe come over to destroy the cordite factory. Searchlights go up from the garrison on Brownsea Island and the first decoy fires are lit on Brownsea and Arne. Old bathtubs packed with wood and coal are ignited. Lavatory cisterns flush paraffin and water down pipes into the burning tubs. With every flush the fires flare up violently in simulation of a bomb blast. Acres of scaffolding, constructed to resemble a giant warehouse, are set alight with tar barrels and paraffin. So it looks, from the air, as if one of the warehouses at Holton Heath has gone up in flames. Everywhere the stench of paraffin. People run to cellars, ditches, cram into cupboards under stairs. One man can't bear the shaking of his garden shelter and decides, though the others in the shelter beg him not to, to sit the raid out in the woods instead.

It is a terrible night on Arne. But the deception works. Five hundred tonnes of bombs are dropped on Arne that night. So many bombs explode on the woods and heaths that the topography of the place is changed. The land is pockmarked, pitted with craters. Two hundred and six bomb craters are counted across the peninsula. And Arne burned and burned. The oak and birch woods burned, the bracken and the heather burned. And the peat burned. Above all, the peat, the underlying structure of the heaths. Once the fire took hold of the peat it would not go out, would not be put out. For weeks Arne smouldered like that. At night planes could see patches of the heath glowing red. The fire brigade would come and put the fire out. Then hours later a rush of oxygen would flush through the peat and stoke another fire into life.

After the raid the people were evacuated. They were given a month's notice and everyone on Arne left the peninsula. A dairy herd on one of the farms was so traumatised by the raid that the cows took sick with milk fever (their milk congealed inside them) and they all had to be destroyed. Four weeks to sell up and get out.

Livestock sent for auction. Boats sold off on the cheap. It was such a hurried thing, the evacuation; they could not even stay for the harvest. Some families went to Blandford, some were given replacement farms elsewhere. Then the army moved in for the duration of the war and Arne was closed off, taken over by the military as a training centre.

After the war a few people came back to live on Arne, but very few of those who left in 1942 returned. The place was in ruins, the village derelict, its houses in an awful state, bullet-chipped, tiles ripped, smothered in ivy. Most of the buildings had to be pulled down. Meanwhile fields had grown rushy, scrub had invaded the ungrazed heath. And hundreds of bomb craters had become pools for dragonflies.

Dawn and dusk. Sometimes it feels like these are the best times to see raptors. You have to get up early for hawks! And so many diurnal birds of prey are surprisingly crepuscular. I have watched buzzards, in particular, hunting through the last dregs of light when it seemed there was barely enough light to fly in, let alone hunt. Hobbies, too, with their large eyes, can hunt for bats and insects deep into the twilight. So I go back out onto the heath in the evening and settle down to wait and watch for hobbies.

A kestrel is here and I watch it above me, quartering the heath. It has young in a pine-tree nest on a slope above a boggy sump. Every time the adult kestrel swoops close to the nest the young call loudly to it with their shrill begging *keeks*. The sound the young kestrels make carries right across the heath. Then, after ten minutes, a falcon's shadow flashes over me. I presume it is the kestrel back again, swooping through another arc of air. But the speed it is going is nothing like a kestrel.

Hobby! At last! Shooting low over the heath, long, sharp wings, a hundred times faster than a kestrel. Hobby: using up every ounce of sky, flinging himself at the air.

This heath is not big enough for him. He circles it in seconds, crosses from one end to the other in a blink. Can he even slow down? Could he go any faster? He is quicker, by far, than any bird I have watched. All the books say the peregrine is the speed master but this hobby feels so much faster than any falcon I have seen. Because he is flying so close to the ground the effect is that the ground is always rushing away from him. He needs to run and run. And then he needs to climb, to fling himself up, only to dive straight back down again to skim the ground. No bird I know flies like this. The closest thing he resembles is a swift, but a hobby is a swift enhanced. Even merlins, those compressed balls of energy, do not burst apart the air like this hobby.

He is hawking for dragonflies. I watch him seize one mid-air. Then, still on the wing, the hobby brings his foot up to his beak. He plucks at the dragonfly, peeling off the chitinous layer, the legs and wings, before he eats the insect. Fragments of dragonfly elytra drop beneath him as he feeds. Then he is on to the next (feeding like this can be frenetic, up to a dozen insects per minute).

A woman from the village has walked out across the heath and has sat down on a bench behind me. I want her to see what I am seeing too. Has she seen the hobby? Surely she cannot miss him, he is mad with speed. I stand up from the heather, grin and point at the hobby. God, what must I look like ... But the woman smiles, waves and her wave seems to say, yes, I am seeing him – witnessing this – too.

What is unique about this experience is that the hobby is not going anywhere. I am not going to lose him suddenly over a ridge or in the depths of a wood. I do not have to try and keep up with him or go looking for him. The heath this evening is his patch. He is not going anywhere else and appears to be feeding for himself, rather than taking prey back to the nest. I just need to sit here and watch as he whirls around me. The only effort in finding him is trying to keep up with his speed.

What is the hobby doing to the air? As he accelerates the air itself accelerates around him, down and back, lifting, thrusting the hobby forwards. In his wake he leaves behind air in mayhem, air like a whirlpool, several whirlpools swirling around each other. He makes a cauldron, a Corryvreckan of the air. He stokes, he concertinas it. The air cannot know what has hit it. If we could see what the hobby does to the air as he smashes through it we would see air in storm, vortexes, geysers, collisions.

* * *

A hobby in its northern breeding quarters is the inverse of its winter self. It is no longer a chaser of rain clouds. The damper western coasts of these islands are not ideal for hobbies; they prefer drier, warmer, insect-alluring habitats. Above all, they are a bird of open spaces, of heathlands, marshlands and farmlands where they have the space to fling themselves after their avian and insect prey. Woods and trees are only for nesting in, these cluttered spaces are the hawks' hunting territory. Hobbies only step into the woods to nest and, crucially, such a wood must provide its own abandoned nests. For hobbies, like all falcons, are squatters, not nest builders. In England, abandoned crows' nests are especially favoured, so much so that hobbies may well depend on a healthy crow population for their own breeding success. You would think that a falcon on its nest would be a no-go zone for other birds. In fact, sometimes the reverse is true. As with geese nesting close to peregrine eyries, so pigeons have been observed to nest in close proximity to hobbies. It is thought that, like the geese, pigeons do this to benefit from the hobbies' fierce protection of their nest sites against predators. Ravens and goshawks, in particular, are abominable to hobbies and the small falcons will harangue these larger predators at every opportunity.

Landscape can be lost, persecuted, reduced as much as any bird, as much as any species. In Dorset only a fraction of the heaths remain which were there in the early nineteenth century. Heathland – that difficult, acidic, nutrient-poor habitat, with its low-growing, woody, oily, inflammable plants (that burned for days on Arne) – has everywhere shrunk. Across southern England, heathland has receded by more than 70 per cent over the last 200 years. Agriculture, afforestation, development have all chipped away at the heaths. What are left are isolated fragments, adrift from one another. What we do to landscape is interrupt it. We plant blocks of conifers across the peat bogs of the Flow Country and cut the bog off from itself, squeezing out the moorland specialists, the golden plover, dunlin and merlin. We flood the Upper Tyne and Upper Tywi valleys with water and forestry until the land no longer recognises itself. Lives, landscapes are interrupted.

If MacGillivray could have afforded to include his watercolours of birds in his *History of British Birds* would that have changed things for him? Would it have changed how his work was received? Would the paintings not have stilled the reviewers who so dismissed the books, their classification scheme, MacGillivray's 'affected' prose, almost everything about the work, even the 'spirit' of the 'Practical Ornithology' sections (his fieldwork notes), which, to my mind, soar with spirit and show how he – William MacGillivray – was one of the finest field naturalists this country has ever known? If MacGillivray had been able to include, for instance, his life-size painting of a grey heron or his watercolour of three linnets perched on a beech sapling, would that not have stopped the critics in their tracks? How might his paintings have rescued him?

The American ornithologist and great bird artist John James Audubon said this of MacGillivray's paintings:

> In short, I think them decidedly the best representations of birds I have ever seen, and have no hesitation in saying, that, should they be engraved in a manner worthy of their excellence, they will form a work not only creditable to you, but surpassing in splendour anything of the kind that Great Britain, or even Europe, has ever produced. Believe me always your sincere friend, John J. Audubon.

Audubon, who leaned on MacGillivray, needed him, needed MacGillivray to help him write his *Ornithological Biography*, the accompanying text to his illustrated *The Birds of America*. Audubon, who came knocking on MacGillivray's door in Edinburgh one day in October 1830 looking for help. And MacGillivray agreed to assist Audubon with his work and correct his manuscripts, his wobbly English and hazy scientific descriptions of the birds for a modest fee of two guineas per sheet of sixteen pages.

The collaboration between Audubon and MacGillivray that ensued was the coming together of an ornithological dream team, the two great American and Scottish ornithologists of the age working together in a haze of bird skins, paper and ink. Audubon starting work early, MacGillivray joining him later in the morning but then working on late into the night. And they worked at such a pace, columns of manuscripts rising up around them. The pair were consumed – swamped – by birds; at night they dreamed of nothing else. They worked like this, on and off, for the next nine years until Audubon's five-volume *Ornithological Biography* had been completed. In the gaps between writing Audubon travelled back and forth between the United States (to procure more bird specimens) and Britain (to write and garner subscriptions for his book). After each field trip to America, Audubon would return to Edinburgh – the engine room – and seek out MacGillivray to resume the writing again.

They would bicker about technicalities, classification, the correct order to arrange the birds. But despite the occasional quarrel and the sheer intensity of the work, the pair became very close friends, working together, walking together, shooting birds together. MacGillivray named one of his sons after Audubon; Audubon named two species of American birds after MacGillivray, *MacGillivray's shore finch* and *MacGillivray's warbler*. Their sons, John MacGillivray and John Audubon, also became friends. John MacGillivray once crashed through the Audubons' Edinburgh home breaking a glass case that housed one of Audubon's birds. If Audubon disapproved of this clumsiness, he and MacGillivray must surely, on another occasion, have secretly approved when their sons were caught poaching together in Ravelston Woods outside Edinburgh and had their guns confiscated by the keeper.

If MacGillivray was not adequately acknowledged by Audubon for his part in the *Biography*'s completion, well, that is MacGillivray for you … self-effacing, not especially concerned with accreditation. But subsequently others have acknowledged that MacGillivray's role was crucial, that without his contribution the *Ornithological Biography* could not have become the great founding work of American ornithology that it is. The American ornithologist Elliott Coues attempted to set the record straight when he wrote of Audubon's *Biography*:

> MacGillivray supplied what was necessary to make his [*Audubon's*] work a contribution to science as well as to art.

But Audubon was not ungrateful. In the entry for *MacGillivray's warbler* in the *Ornithological Biography*, Audubon wrote:

> I cannot do better than dedicate this pretty little bird to my excellent friend William MacGillivray Esq. I feel much pleasure in introducing it to the notice of the ornithological world, under a name which I trust will endure as long as the species itself.

Still, MacGillivray's name has not endured. He has slipped from view, his books and paintings have slipped from view. And *MacGillivray's shore finch* was later renamed the *seaside sparrow*. Audubon, by contrast, is renowned the world over and his paintings sell for vast sums today.

In his introduction to volume V of the *Ornithological Biography*, Audubon wrote:

> Allow me also to mention the names of a few friends to whom I shall ever feel most deeply indebted. The first on the list is William MacGillivray, and I wish you, Reader, and all the world besides, knew him as well as I do.

I envy Audubon greatly, that he knew William MacGillivray so well.

The hobby is still here, hunting over the heath. Fast and low, charging straight up, then down again to skim the heather. Sometimes he hangs a while in the air not unlike a kestrel. But never for long and he is always off again running into the wind. As the sun drops lower I see the hobby more sharply: white face, black hood, grey-black dorsal, a faint rusty orange high up on the underside of the tail and around his legs. Every time he turns into the low sun I can see the mottled white patterns of his breast and the reds below his chest.

When the hobby passes close to the young kestrels in their pine-tree nest, the kestrels call to him. They must confuse him for their parent, that sharp-winged falcon shape. They are roughly – kestrel and hobby – a similar size, though the hobby appears a little smaller when he perches briefly on a fence post. He seems to shrink into himself then, much as I've seen sparrowhawks do when they are also perched.

I clamber out of the heather and walk over to the woman still sitting on the bench behind me. *Fast as a peregrine!* is the first thing she says to me. She has lived in the village for fifteen years, she says, and has never seen anything as impressive as this hobby's display. We stand there together for a while in the middle of the heath. Then the woman is turning to head for home, *Goodbye, well that was very special, wasn't it ...*

I don't want to leave. I think the hobby is still out there, but the light is hopeless now. The last thing I see before I go is an egret flying slowly up from the shore, crossing the heath in front of me, heading for the wood to roost. A slow, heron-heavy flight, porcelain white. As the egret passes in front of the pine trees that skirt the heath, the bird suddenly brightens, becoming whiter against the dark backdrop of the pines.

XIV

Buzzard

River Teign, Devon

MacGillivray leaves Buxton on the Friday at twelve o'clock. It is one of those rare days when he is immune from melancholy. Crossing the high hills of Derbyshire he feels like he is soaring there and, for the first time on his marathon, he senses he is closing in on London.

> Buxton to Derby: 34 miles
> Derby to Leicester: 26 miles
> Leicester to Northampton: 32 miles
> Northampton to London: 66 miles

Up on the hills, that day, he sees: skylark, rook, starling, hedge sparrow, pied wagtail, wren, chaffinch, magpie, yellowhammer.

That night, in a scruffy, smoky lodging house outside Ashbourne, MacGillivray sits up late, smoking, comparing Scots Gaelic to Irish Gaelic with an Irish pedlar, then trying out his Latin on a young Italian salesman to see how the Italian and Latin words relate to one another. How these sister languages fit inside each other is not dissimilar to the ways that different species of birds coincide and resemble one another. Through his dissections, his close observations of anatomy, plumage, diet and behaviour, MacGillivray tries to see how the birds converge and form a passage into one another. In *Descriptions of the Rapacious Birds* he writes:

> The genera Circaetus and Harpyia connect the eagles and osprey with the buzzards, while the Morphni would seem to unite them with the asturs. But of the birds that occur in our country, none are intermediate between the eagles and buzzards.

The next morning, two miles from Ashbourne, the musk mallow is in flower, and further on MacGillivray finds the ivy-leaved toadflax growing on a bridge. In Derby he buys bread and cheese and eats it in a field where water figwort is showing. He leaves Loughborough at three o'clock and the air is sharp with frost. There are vines growing up the sides of houses with tiny embryonic grapes. Field-mouse-ear chickweed, black horehound and common creeping cinquefoil all delay him. It is a day of sudden bursts of light: a glint of green beside a stream is a kingfisher and, in the dusk, as MacGillivray passes through Leicester, he sees, briefly, the Aurora borealis dancing, lighting the clouds to the north. Later it begins to rain and he finds a barn and buries himself in a heap of straw with his knapsack for a pillow. But it is far too cold to sleep and around two o'clock he gets up and continues walking. The dark propels him as if he had somehow grown much lighter. Figures loom towards him through the darkness and he flinches as they pass, but nobody tries to hassle him and he seems unaware how intimidating he must have seemed to others on the road, what with bits of straw still sticking to him and his knapsack swelling – doubling – his size in the dark. At first light the land is under frost and MacGillivray shoves his hands (he lost his gloves three weeks ago) deep into his pockets to muffle them. His feet ache and he is dizzy with weakness. The next milestone is the 80th from London.

Bread for breakfast and, cutting it with his knife, MacGillivray slices his thumb to the bone. Near Northampton he finds specimens of common nightshade, greater knapweed, dwarf mallow. And that evening, despite walking 51 miles without sleeping, he is able to write this wonderful sentence in his journal:

> Flora still continued to smile upon me and in the evening I found the Common Traveller's Joy which I examined in the Inn.

The inn is at Grafton Regis between Northampton and Stoney Stratford. That night MacGillivray sleeps under warm blankets and wakes at nine. Fifty-eight miles to London and he only has thirteen pence halfpenny left. So onwards without eating and the hard road is gnawing at his feet. His shoes and stockings have dismantled themselves and he has become such a ragged, hobbling thing. He buys some bread to chew and by midnight he is passing through St Albans. He lies down under a tree and tries to grab a few hours' rest, but then he is up again and crawling on – cocks crowing – a herd of oxen swaying down the road – still 18 miles to London – heavy rain – threepence halfpenny remains, he can hear the coppers sloshing in his pocket – a bite of bread – an apple – god, how his feet ache – how his thumb aches – his clothes are soaking – it is still raining – his long blue coat is heavy with rain – soon after midday MacGillivray is staggering through Highgate – 838 miles since he left Aberdeen – his knapsack, his 'machine', so stuffed with specimens of plants it looks like he has grown some strange herbal appendage on his back ... Jolting, grimacing, dreaming the final mile into London.

Buzzard

The upper and fore part of the cere is bare, but its sides are covered with bristly feathers, which are downy at the base; the space between the bill and eye is pretty closely covered with radiating bristly feathers, slightly downy at the base; the sharp projecting eyebrow bare; the edges of the eyelids furnished with ciliary bristles. The plumage in general is full and soft, but rather compact and glossy above, although the margins are loose. The feathers of the head are small and narrow, those of the neck larger and more rounded, of the other parts broad and rounded; but the plumage has not, as some allege, any decided resemblance to that of the owls, being, in fact, as firm as that of the goshawk, and, in proportion to the size of the birds, as that of the eagles.

Coming down the lane in the car through blue June warmth. Just before we splash through the ford I see the buzzard.

– Boys, look up!
– What is it, Dad?
– A buzzard, can you see it?

I stop the car.

A commotion in the high branches of an oak. The leaves obscure what is happening but there is a screech and flash of magpie. Then a glimpse of buzzard, 20 feet away, trying to untangle something from the tree.

– What is it, Dad?
– Is it an eagle?
– There it is!

We see the buzzard detaching itself from the branches and the thing dangling from its talons gives the buzzard a different profile in the air: a limp, black shape only identifiable as a fledgling magpie because the two adult magpies are in the field beneath the tree, bobbing, yikkering with distress.

Buzzard
The bill is black, near the cere greyish-blue, its soft edges yellow; the cere and bare space over the eye greenish-yellow; the irides yellowish-brown; the feet bright yellow, the claws like the bill.

From Dorset I turned west then south to Dartmoor. I wanted to walk along a stretch of the river Teign and follow the river from its source high up on the moor down through its wooded, buzzard-rich valley to the east.

I reached Chagford in the early afternoon and set off walking over Meldon Hill to the south of the village, weaving through banks of gorse, ponies parked in amongst the gorse, stubs of granite around the summit. Coming down off the hill into deep, high-banked lanes; sunken, subterranean pathways, you cannot see out of them until you pass a gateway cut into the side of the hedge. It's only then you catch a glimpse into one of the small fields that flank the lanes. Those fields seemed so remote, enclosed by their tall, thick hedgerows. In one of the fields I glimpsed, through its gateway, a flock of rooks and jackdaws, the jackdaws noisy in their squeaky chatter, bobbing around the larger rooks.

Then, climbing out of the lanes, approaching the moor, and up ahead a buzzard rises from the verge. It grows in flight! Huge, broad, long-fingered wings unfolding. It lands a little further up the road on a wooden telegraph pole and I sit down on the bank to watch. The dark creosoted pole accentuates the bird's greys and whites. Its grey face twitches and bobs, scanning the field in front. Then it is dropping from its perch in a low glide. Lands in a hunched pounce, talons first, stabbing at something in the grass. Misses: is quickly up again, this time landing on another pole further up the road.

Just as the kestrel hangs inside the air – hovers there – the buzzard, so often, hunts statically, scanning the ground from its perch in a tree, on a pole or fence post, waiting, listening for a vole to crinkle the grass below. And so often, as with so many British raptors, it is field voles (short-tailed voles) that a buzzard is hunting. Voles (particularly during the winter months) are a staple of the common buzzard's diet. If not voles, then: rabbits, insects, moles, frogs, earthworms, carrion, road-kill and, during the breeding season, fledgling birds. Insects (grubs and beetles especially) are also consumed in large quantities by buzzards. As are rabbits, notably in spring and early summer, when buzzards need to hunt through all the long daylight hours to provide for their young.

Rabbits and voles are to the common buzzard what wasps are to the honey buzzard. Without these crucial prey species the common buzzard would struggle to rear its young successfully.

The buzzard is both a static hunter, a hoverer and a low-level searcher. Sometimes I have seen buzzards poised in strong wind like a kestrel, their carpal joints pushed forward, their wings braced back so that the bird assumes a tight, streamlined shape, nothing like the broad, upcurved wings and fanned tail of a soaring buzzard. Other times I have watched buzzards hunting low and fast across the ground much like a golden eagle. And seen them too hunting like a hawk, swerving fast through woodland. The buzzard seems capable of taking on the hunting guise of many different birds of prey.

High-up soaring buzzards are not hunting. They are displaying, marking their territory, making the most of a rising thermal. The smaller, lighter male can rise more quickly than the female; often, when you see a pair of soaring buzzards, they are layered, one above the other, with the male invariably the higher bird. When the pair come together, as for example when the male swoops down at the female in his spectacular display flight, the difference in size between the sexes is clear. Soaring buzzards often call to one another and their calling will sometimes draw in other buzzards from adjoining areas to join their circling flight. On a still day their calls travel far. Cliffs ricochet a buzzard's call, making it sound sharper; woods soften and dampen the call; valleys carry it along their alleyways, extending it. A buzzard's call is among the most beautiful sounds I know.

A hunting buzzard is altogether different, capable, despite the buzzard's reputation for sluggishness, of astonishing bursts of speed. Many times I have been taken aback by a buzzard's turn of speed and on several occasions have wished-mistaken a buzzard into a hawk because it was flying so fast through the trees. Once, in a wood near my home, a bird rushed past me just 10 feet away and landed on the long thick branch of an oak. In the moment between it brushing past me and landing, before I could focus my binoculars, I was convinced, because of the speed with which it had shot into the wood, and because of the bird's size, it must have been a goshawk. But no, it was a buzzard, an adult, who, as soon as he touched down on the branch, was joined by a juvenile. The adult bird dropped some prey on the branch then quickly left. The young bird picked up the prey item in its beak, swallowing it whole.

Buzzards will also hunt on foot and, even when on the ground, can surprise you with their speed. They will graze through a field on damp early mornings for earthworms, or stand beside a mole run and wait for the earth to quiver, sprinting across to grab the mole where it surfaces.

I sit on the bank for half an hour watching the buzzard hunting from its perch on the telegraph poles. Head bobbing, judging the distance and angle of its swoop. Then: launch, glide, pounce, stabbing at something in the grass before returning to its lookout on one of the poles. On it goes like this. When the buzzard finally flies off, I get up and follow the small road west into a large conifer plantation. The wood is dripping, glittering from a recent heavy shower. The rocks and trunks are coated in moss; in places, where the light reaches the wood's understory, the moss emits a brightness, a green luminescence. Then I am out of the wood and dropping down a slope and I can see the headwaters of the North Teign in front of me, deep-pooled, fast and black. I walk down to the river, pick it up like a path, and follow it up onto the moor.

South-west, due west, north-west: erratic thing you are up here, river, trying to figure out a way off this vast granite plateau. I am looking for a place to camp the night near the river's source when I see another buzzard, querying the moor grass, hunting low like a harrier. And for a moment I mistake it for a harrier until I can see the bird is definitely a buzzard, a shape-shifter, flying much lower than I'm used to seeing, but the broadness of its wings – the sheer bulk of the bird – gives it away.

An hour later the moor is cooling fast and in the dusk light, sitting outside my tent, I have a wonderful, unexpected visitor. A merlin lands on the stone dyke just 12 feet away. If, that is, you can say a merlin ever really lands, ever settles properly. It is a bird that is always on the point of leaving, a quivering pause before it springs off again. Sharp-pointed wings, forked behind her like a swallow. Hello merlin, how good to see your shape again … Please don't go just yet … But when I shift my binoculars a fraction she is off, fast and low over the moor. Hello-goodbye-gone! Messenger from Orkney and The Flows and the winter marshes of Sheppey. Almost at the end of my journey and what a gift to have it linked up like this, a line drawn by a merlin from Orkney all the way to Devon.

In the morning: thick, cold mist. I can see only a few yards around my tent. The end of September and the high moor is already in another season. The moor has drifted so far ahead into autumn it has become inexplicable to the land – the *in-country* – below. A pony comes out of the mist like an apparition and wanders through the reeds in front of me. Its mane is beaded with dew, its coat is a tinny, bracken-red colour. I will not see any birds until this mist has burnt off, so I strike camp and walk on up the river to try to find its source.

London is: MacGillivray getting out of his drenched clothes; his feet exhaling; collapsing at the home of Mr Cowie, 31 Poultry (whose address MacGillivray had been given by his friend, Dr Barclay); washing his feet with warm water; opening a letter from his aunt, Mary MacAskill from the Isles; his feet are barely recognisable, they are so pale and crinkled they seem mummified; 838 miles since he left Aberdeen; *what a queer sort of dream this journey of mine has been*; then sleep and sleep and waking at nine into (oh bliss!) fresh clothes.

Before he heads out to explore the city he remembers he has forgotten to mention in his journal the plants he examined on the outskirts of London. And I am amazed in the blur and agony of those final miles he had the wherewithal to even notice these things:

> I forgot to mention in its proper place that I had examined on Tuesday the Blue Bramble Rubus caesius. It occurred in several places by the road, and I eat a great quantity of its berries which I find delicious and quite unlike those of the Common Bramble Rubus fruticosus which are most nauseous. Yesterday I found and examined another plant by the road near the fifteenth milestone, the Wild Succory. The Small Bindweed I found very common, but it is out of flower. I have now finished my journey and I am satisfied with my conduct.

London is: snowing heavily on the night of 22 October 1819; MacGillivray walking everywhere; climbing the Monument and peering out across the city through the smoky air; paying three shillings to visit the sad collection of animals at the Exeter Change menagerie; stopping to listen to a young girl singing in the street; dipping inside Westminster Abbey and MacGillivray coming out grumpy and angry at all the monuments to generals and admirals (what about the scientists and poets, he cries!).

Monday 25 October, 1819. MacGillivray is one of the first through the doors of the British Museum when it opens its rooms at eleven o'clock. He gravitates past the displays of minerals and shells towards the collection of British Birds, which has a room all to its own. What happens to MacGillivray in that room? He has walked all the way from Aberdeen to here, for this. It is an important moment. For ornithology, as significant, surely, as MacGillivray's meeting with Audubon. What is important about the time MacGillivray spends drifting through the collection is that it gives him confidence. He realises that his own self-taught theories of ornithology are accurate, that his intuition about these things is correct. That knowledge gladdens him, warms him inestimably. The ways the birds have been arranged according to their genera are much as he would have arranged them (though he also feels he could improve on the museum's system). What is most significant about his visit to the museum is that MacGillivray comes away from it knowing for certain that this is what he must do with his life. He *must* study the natural world but, above all, he knows that this approach – the sterility of the museum, all those specimens locked behind their glass cases – that approach is not for him. If he is to study the natural world, it must be out there – *he* needs to be out there – immersed, on the interface with nature:

> I felt my love of Natural History very much increased by the inspection of the museum. At the same time I felt convinced that to study nature I must have recourse to nature alone, pure and free from human interference.

What MacGillivray writes in the closing pages of his journal, he does not need to. Already – for some time now – he has achieved what he aspires to:

> Ornithology is my favourite study and it will go hard with me if I do not one day merit the name of ornithologist, aye, and of Botanist too – and moreover of something else of greater importance than either.

The mist is clearing above Whitehorse Leat. The first inkling, a thinning where the sun is a faint laser trying to burn through. Then, quite rapidly, the mist is being disassembled and I can see a tor's rubbled top, its clitter spill. Then I see the sharp spruce edge of Fernworthy running down the long back of White Ridge.

There is a kestrel over the ruins of Teignhead farm. Below the kestrel, downstream beyond the clapper bridge, a buzzard is rising. Though he may take on the hunting guise of so many different birds of prey, this morning, soaring above the river, the buzzard is himself, calling his high echoing call that is unlike any other raptor's. Sometimes I think I can detect a *w* or *p* sound in a buzzard's call: *wee-ooe* or *pee-ooe*. The first syllable usually short, followed quickly by a longer, wavering *ooe* sound. Other times, though I try to decipher it, I cannot hear any 'letters'. It is just two high notes with the second note sounding – feeling like – a reverberation of the first. It is always unmistakably a buzzard but the call (that second note especially) can vary in length and pitch. Sometimes that note is left to carry, other times it is brought up short and raised in pitch. Occasionally the first note is so short-lived the call feels like one long note, rather than two in quick succession. In late summer, while they are still in the vicinity of the nest and being fed by their parents, juvenile buzzards call a great deal and their call is distinctive, sounding longer, sharper, higher-pitched than the adult bird's. This morning, above the moor, the buzzard's call lingers on and on with an echo's resonance as if the bird is testing – using its voice like sonar – to gauge the depth of the valley.

Buzzards call in aggression, agitation, in courtship, to ward and warn off other buzzards. They are tenacious in their defence of territory and pursuits and squabbles with trespassing buzzards are vociferous affairs. Kestrels call often too with their high trilling whistle, but I don't think any other bird of prey is as conspicuous by its call as the common buzzard. Often you will hear a buzzard before you see it. How different, in this respect, the buzzard is to the largely silent golden eagle. I have watched buzzards flanked and jostled by a mobbing raven, the buzzards calling in what sounds like annoyance or frustration. The call always sounding shorter, more agitated, when the buzzard is being mobbed like this. Most of the time a buzzard's calling is prompted – is a part of – the bird's territorial display. A buzzard calling while it soars above its patch is like an acoustic 'beating of the bounds' of its parish; the long-drawn-out calls carry to the boundaries where they reverberate then settle.

Cattle are out along the side of Great Varracombe. I walk through them as I climb the bank up to Teignhead farm. Campfire pits in amongst the ruins, their ashes congealed to a black mulch, rain-prints dent the ash. A small shed with a rusty, sphagnum-coloured, corrugated-iron roof. The walls of the old farmhouse are huge, thick slabs of granite. How many ruins have I rummaged through on this journey? That abandoned house far out on the Caithness flows; the empty shell of Coventry cathedral; the chequered black and white floor tiles that were all that remained of Lord Leverhulme's home on the side of the hill above Rivington; the view from the open window of the abandoned croft on Oronsay across to the hills of Ardnamurchan ... It is strange how a ruin can pull you in, compel you towards it from miles away.

What happened here at Teignhead? The farm's buildings were substantial: a large courtyard with its own walled-in dung-pit, a sheep-dip, paved paths leading out to the moor, clapper bridges laid across the brook and river ... A leat is cut to carry water from the hill to a well outside the front door. Gateposts are quarried from the rock on Magna Hill and left there until a fall of snow allowed them to be strapped to sleighs and sledded down to the farm. Seventeen eighty, the first tenant moves in. Sometimes the farm is cut off by snow for weeks. But even more than snow, it is the mists that cut the place off, stubborn, heavy mists that will not shift, that snag over the farm for days on end. It is as if the moorland bogs somehow engineer the mists, breathe them into being. By 1808, 1,551 acres of the farm have been enclosed and the long stone walls that have been built across the moor (one of these is where I had seen the merlin) are causing offence to the Gidleigh commoners who feel their rights and access to the common grazing lands are being encroached. Eighteen seventy-six, the incumbent tenant is evicted by the landowner, the Duchy of Cornwall, because the tenant's rent is in arrears. Nineteen forty-two, the land is requisitioned by the War Office and troops move in for training exercises. The following year, 1943, the last tenant leaves Teignhead farm (bundled out with some compensation from the War Office) and goes to live with a relative in Chagford. As happened at Arne during the war, the military take over the area completely. But unlike at Arne, after the war was over and the army had left, nobody moves back to Teignhead. By 1950 the place is in a bad state, the enclosure walls have begun to split, the buildings disfigured by the wind and vandals working off each other.

What happened at Teignhead is what happened to Leverhulme's summer home on the moor above Rivington. It is the same trajectory, the same story. Encroachment is followed by retraction, and before you know it the place is crumbling and the moor has crept back in again. What happened to Teignhead is also what happened to the hill farms of the Upper Tyne and Upper Tywi. There is an inevitability in the way it ends for these upland farms, as if these places endure a different level of gravity, as if something is constantly pushing down on them so that in the end it becomes impossible to cling on up there.

Dartmoor has always been drawn over, sculpted, its surface etched with standing stones and monoliths. Its granite outcrops are beautifully irregular and the wind like a potter continues to work at them. More recently, the peat (no longer accumulating on the high blanket bogs because of climatic shifts) has begun to shrink. The ensuing erosion looks as if someone has torn chunks out of the ground, pools and canyons weaving around islets of peat.

In the nineteenth century a great redrawing of the moor took place. Stone dykes were erected and charged off into the distance. Rapidly the moor was repatterned, made linear. On Dartmoor the term for this newly enclosed land was 'newtake' and by the end of that century the northern part of the moor had been completely severed from the southern by newtake walls. Inevitably, there were disputes: occasionally you still come across walls on Dartmoor that suddenly peter out, where the commoners had managed to assert their grazing rights and halt the enclosure. Besides these enclosures made by the larger tenant farmers (as at Teignhead), many small enclosures were made by the commoners themselves, appropriating, parcelling off tiny pockets around the edge of the moor. Often these small enclosures were thrown up in a single night, claimed and asserted in the same way that the 'one-night houses' in the marginal lands of rural Wales were.

Debatable, Batable, Disputed ground ... Every seven years on Dartmoor parishioners go out to beat the parish bounds. The boundaries themselves are often hazy, fluid, contentious things. It is not so clear-cut where the moor (the Forest) and Common divide. And boundaries can be porous, shifting lines. So every seven years people from the parish walk out along these boundaries, repairing, extending, adjusting the lines. Traditionally it is an exercise conducted in a spirit of subversion and the purpose, wherever possible, is to infringe anything imposed by the landowner. As they circumnavigate the parish, whenever the party arrive at a significant boundary marker, the custom is for the oldest man in the group to pick up the youngest boy, flip him upside down, and tap the youngster's head on the boundary marker to ensure that the younger generation stores, and does not forget, the position of the border.

MacGillivray wrote:

> A river is nothing but a continuous series of continually renewed drops of water following each other in a groove.

But when I drop down off the moor and pick the river up again in the shade of the woods, it is unrecognisable. All that moorland energy has dissipated, it has settled into itself. It has become a river of deep pools and sandy beds, a hoarder of leaves and branches, tucking them under its rocks to store there. And fish hang in its pools like shards of mica.

For the rest of the day I follow the river eastwards. Often it does not want to be seen, hidden under dark banks of rhododendron, contour-screened. And that is fine, I am just as happy to follow its sound-road, to veer away but to stay within earshot. The path deepens as it drops down from the moor, becomes a cutting, almost a tunnel. Holly and gorse line the high banks on either side. Badger runs criss-cross the path, I can see their claw marks in the soil where the bank is steepest. A buzzard drops from a tree and swoops low and fast through a narrow gap in the hedge. Its face is a light grey colour; wings and back a sandy brown. Again, I am struck by the huge bird's agility and turn of speed. A tussle through the hedge-gap with a crow: then gone.

The way frost peels and recedes across a field until there is just a small pocket, a corner of the field, which the low winter sun cannot touch and where the frost sleeps all day, that is what happened to buzzards in the nineteenth century. They were exterminated from lowland Britain, peeled back, like the red kite (and for the same reasons as the kite), to the corners, the fringes of these islands. Devon and Cornwall, the New Forest, parts of Wales, the West of Scotland and North West England held remnant populations of buzzards. That the buzzard has crept back across most of the rest of the country (becoming our most common bird of prey), spreading east, colonising old lowland haunts, is a remarkable turnaround, explained by a new era of tolerance and by the buzzard's adaptability, its versatility to both a variety of habitats and a variety of prey.

Bird of woodland, bird of farmland, bird of moorland, wherever these places meet and mix. The ideal buzzard habitat is one which contains a bit of each. Mixed, irregular farmland is ideal. A monochrome landscape is not so good, at least when it comes to breeding success. The buzzards I have watched closely hunt especially over the brambly, rabbity, vole-rich, unkempt no-man's-lands, the steep banks and field edges that are so important to so many birds of prey.

And buzzards, too, will often lead you to other birds, often other birds of prey. They are such a conspicuous, vocal bird. When you see a buzzard it frequently draws your eye to something else that has been drawn in by the large bird's wake. I have watched a sparrowhawk displaying in early spring beside a rising buzzard, seen kestrels and ravens and other corvids (especially ravens) sparring with buzzards. Throughout this journey buzzards have been a helpful guide, more often than not standing in for other raptors I have been searching for, but sometimes, too, helping me to find these other birds.

Downstream, in deep woods, the valley narrowed to a gorge. I'm trying to make out the mix of leaves gyrating slowly, a slack flotilla in the river's current: beech and birch, maple, alder, oak, some pheasant feathers in amongst the leaves. What is flushing the pheasant poults out of the trees and across the river? Something is panicking them. There is a Coke can floating in the shallows. I wade out to retrieve it and the water is lovely and cool on my moor-weary feet. An otter has left its tarry spraint, bitty with tiny bones, on a mossy boulder. A buzzard has started calling somewhere above the river.

To be stilled – stopped – that is what birds of prey do to me. There is a phrase particular to Dartmoor: *The Ammil*. It describes the phenomenon, occasionally witnessed on the moor in winter, when everything exposed to the air, every blade of grass, every rock, every twig, becomes encased in ice. The Ammil is the thaw put on hold, paused for a while. The temperature suddenly drops below freezing and the thawing, dripping, running world is held in check, suspended, so that everything, even the great rocky outcrops of the tors, is sheathed in ice. *Ammil*: from *ammel*, the Old English term for enamel, to encrust, to coat with a vitreous sheen. A rare event. The moor is decorated, it glistens. The land is stilled, and to witness it is also to be stilled.

XV

Sparrowhawk

Home

Home is: three doors down from the end of a street on the edge of a small village. The street ends at a kissing gate. The gate leads through to a grassy field, boggy around the gate, rising quickly to a steep bank. Hawthorn, elder, gorse and bramble grow along the bank, sporadic oak and ash trees too. Bracken dominates the western end. In May, before the bracken takes over, there are thousands of bluebells. I love the range of colours along the slope, there always seems to be a brightness there: the elder in flower, hawthorn blossom, the pink rosebay willowherb. Even in winter there is yellow on the gorse. Foxes have their den deep within the gorse. Ravens nest in a Scots pine windbreak above the slope. At the eastern end, where the trees thicken to a spinney, there is an ancient badger city, a lumpy quarried place. Roe and muntjac use the bank for cover. Fallow make channels through the bracken in summer. From the field – from the washing line in my back garden – the top of the bank forms the horizon. Sometimes a swollen moon seems to snag itself on the bank as its huge pale dome rolls along the horizon. The soil is heavy clay. The field is full of wounds, the slightest rain sets it weeping.

To the west of the field there is a large wood. A medieval bank marks its edge. Ancient ash stools, like eroded gargoyles, line the bank. Dog-rose and bramble form a hedge between the wood and field. The wood stretches west for nearly a mile and goes through different belts of trees: larch, spruce, sweet chestnut, the occasional colossal beech, its trunk black with rain seep. But here, in the corner closest to the field, it is mostly oak with hazel coppice. Ash grows around the wood's edge and there are some dark pools of holly in amongst the hazel understory.

For many months I walked all over the wood and further into the fields beyond looking for sparrowhawks. I walked miles searching for the birds. But my technique was all wrong. They are like ghosts, sparrowhawks, they cannot be tracked or followed; they appear from nowhere in a rush of speed and then they are gone. Sparrowhawks operate in a different dimension from other birds of prey. They live in the pocket of air that hugs the surface of the land. Quick and low, they cut the ground like a scythe.

Still, I pursued my wanderings through the wood and fields. Sometimes I came across a hawk's plucking post and occasionally – more rarely – there was a snatched glimpse of hawk disappearing through the trees. In deep summer, when the wood droned with the sound of hoverflies, I often heard the sharp begging calls of young sparrowhawks. All these things encouraged me into thinking I was getting closer to the birds. But that was never true: with every rare encounter the hawks would briefly shimmer then dissolve. Always, though, the encounters, when they happened, were breathtaking. Once when I was walking down the side of a field towards a brook there was a movement 10 feet away. Rising out behind a bramble thicket beside a gate was a female sparrowhawk, wings spread and tail opened in a fan as she rose. I could see the dark striped bars running across the underside of her tail. As she climbed above the thicket she dropped something, a grey shape. I waded through the brambles and there, next to the gate, was the body of a pigeon, head missing, a deep red gash down its breast.

Most encounters happened when I was not looking for the birds. Early one morning, putting a bag in the boot of the car, the quiet street suddenly interrupted by a clatter of pigeons and in their wake a sparrowhawk sheering over the kissing gate. I would sometimes see sparrow-hawks when I was driving slowly through the village. Once I pulled the car up beside a hawk I had seen land on a neighbour's wall, so close I could see the speckled patterning on her breast and the yellow flash of her eye. Another time – an astonishing encounter – a hawk flashed low in front of my car and slid over a garden gate. The hawk looked like it was going to collide with the front of the house, but at the last moment it shot straight up, glanced over the roof, and dropped down into the back garden. I had never seen any bird steer itself past objects at such speed. No bird I know flies this fast and low.

Sometimes garden birds would warn me of a hawk approaching, chaffinches and great tits at a garden feeder suddenly ratcheting up their anxious chatter. Sure enough, three seconds later, there was a male hawk swooping across the lawn and, as he rose to clear a hedge, I could see the rusty orange flecks that lit his breast.

In early spring the washing line is a good place to watch. On a fine day in March I sometimes see sparrow-hawks displaying above the bank. My elderly neighbour often pauses by the fence to chat while I hang the washing out. I like to linger there as much to talk to him as watch for birds. Sometimes he brings offerings from his garden: potatoes, small thick cucumbers, delicious red and yellow tomatoes. Grass snakes lay their eggs beneath the empty plastic compost bags in his greenhouse. He walks slowly and explains his hip is no good since a shire horse backed into him when he was young as he was leading the horse through a narrow gate. His anecdotes animate the field, the bank and the wood beyond. Tanks, he tells me, were parked at the bottom of the garden during the war and tested on the bank's steep gradient. The army camped in the wood and bartered sugar and tea for vegetables from his parents' garden. He remembers a Wellington crashing on the bank and he and his classmates rushing out of school and up the hill to find the plane on fire. I ask if he remembers the bombers coming over on their way to Coventry. He nods and tells me that there used to be a farmer who kept a few chickens up on the hill above the village. One night during the war a German bomber returning from a raid (and needing to jettison an unused bomb) must have seen the light from the farmer's torch glittering as he locked the chickens up. *Missed by a mile, thank God, but for a long time you could see the hole made by the bomb up there.*

Often, while we chat, buzzards run the gauntlet of the raven's airspace above the bank. A raven's loud, rapid croaking usually indicates a buzzard has appeared. Recently red kites have settled in the area and I see them almost daily now, their ash-grey heads conspicuous as the birds perch on top of a bare winter ash. I doubt these birds have been seen over the wood for two hundred years. Very occasionally I hear one calling, a high thin whistle, thinner than a buzzard's call. The kites appear taller than the buzzards when they perch, their long folded wings give them a longer back. Perched and silhouetted against the sky, the kites can seem huge; clear winter skies seem to magnify them.

William MacGillivray's home on Harris is a ruin now, the buildings from his uncle's farm commandeered for a sheep fank, a maze of gates, walkways and holding pens. The floor of the fank is littered with scraps of wool. The ruin sits on the lower slopes of Ceapabhal, looking east across the great expanse of Scarista Sands and north to the mountains of Harris. This was MacGillivray's childhood home and somewhere on Ceapabhal's rocky neck was the pit where he concealed himself to shoot the golden eagle. The low walls of the ruin are blotched with amber-coloured lichen, nettles grow in the cracks between the stones. The lush machair runs down from the ruin to the beach, its seed-heavy grasses pricked with yellow, white and purple flowers: eyebright, red clover, birdsfoot trefoil … A smaller ruin, 40 feet by 12, stands to the side of the old steadings. It is being used to store coils of fencing wire. Sections of its walls have been patched with mortar. Meadow pipits, with their streaked breasts, flicker along the top of the wall. The way the pipits blend into the backdrop of the stones, when the birds skitter about the ruin, it looks as if the wall itself is trembling.

From MacGillivray's Hebridean journal:

> **Monday 10th November, 1817**
> Today I rose early, drank warm milk at the gate as yesterday – then walked along the shore of Tastir,

over Traigh-na-clibhadh, and along the rocks to the upper end of Moll-na-h-Uidhadh, then crossed Ui to the great sand, and returned along its margin. In this course the birds seen were the Starling, among the cattle and in the corn-yard, the Shag on the coast, the Common Gull in South town and on Ui, the Great Black-backed Gull on Moll-na-h'Ui, the Curlew on Ui in large flocks, the Wren on the marsh dyke of Ui, the Meadow Pipit on the shore, the Hooded Crow on Ui, the Raven on Fastir, and the Ringed Plover on the sands …

I have determined to describe all the birds found in Harris & shall fall to work immediately. In the evening I went to Moll to search the shores for shell-fish …

Wednesday 12th November, 1817

… The description of birds should be made in the following order. Name, (Linnaean, English, Gaelic) Description proper, very minute. Bill, feet, irides, general colour, dorsal, sternal, colours, habitation, migration, nuptials, nidification, ovation, incubation, education, food, use. Bill, dimensions, colour, shape, nostrils, tongue, shape, colour, mouth colour, eyes, appendages – feet, legs, shape, colour &c. toes number, nails.

Monday, 27th April, 1818

… I have not yet seen an account of the Birds of Britain with which I am entirely pleased; and I have

> of late been thinking upon the subject. Perhaps it might not be a mad scheme to attempt the Ornithology of Scotland. I certainly would not engage for more. But whether this alone would be acceptable I cannot yet determine. However I shall begin to note every particular regarding it, which I can observe, or collect from creditable authority. The time I occupy in this will not be misspent, even at the worst. For I will thus perhaps acquire habits of attention, observation, and activity.

After his visit to the British Museum, MacGillivray left London on 28 October, returning to Aberdeen by sea. He did not have enough money for a cabin and took passage instead in the misery of the ship's forecastle. Perhaps someone took pity on him there or he pleaded a discount, for after two uncomfortable days he was upgraded to a cabin and arrived back in Aberdeen on Saturday 6 November, 1819. The return journey by sea was about 450 miles. In all, since he walked out of his front door in Aberdeen two months earlier, MacGillivray had covered 1,288 miles.

> From Aberdeen to Gretna: 501 miles
> From Gretna to London: 337 miles
> 838 miles
> From London to Aberdeen by sea: about 450 miles
> In all: 1,288 miles

The last entry he writes in his journal:

> My journey is now finished. I arrived here early on Saturday the 6th after a disagreeable passage of ten days from London in the smack Expert. I am again plunged into the gulf of actual existence, and I can scarcely brandish a quill. Sapient remarks and practical conclusions, and resolutions sine fine should now be made, but all these fine things I must defer till I have another journey …
>
> Vanity of vanities – all is vanity. What profit hath a man of all his labour which he taketh under the sun? For he that increaseth knowledge increaseth sorrow.

For the next few years, after the walk to London, it is hard to keep track of MacGillivray. There are hints here and there of his whereabouts. The plant specimens he collected, carefully labelled with a date and location, offer clues as to his wanderings. Scotland in the 1820s becomes triangular for him and he seems to have travelled regularly between the three points of that triangle: Harris, Aberdeen and Edinburgh. Harris is where he hones his fieldwork, where he *hammered at the gneiss rocks, gathered gulls' eggs, and shot plovers and pigeons* … and where, in September 1820, he marries his aunt's younger stepsister, Marion McCaskill.

Edinburgh in the 1820s is where he has not yet settled into himself. He attends the lectures of Professor Jameson, who held the Regius Chair of Natural History at Edinburgh University. And Jameson must have seen something in MacGillivray, something glinting there, because MacGillivray becomes Jameson's secretary and assistant in 1823 and he is also put in charge of Jameson's substantial Museum of Natural History. A dream job, a kid in a candy store: the museum held specimens of birds, rocks and minerals in their thousands. And for a few years MacGillivray is pin-down-able, working at the museum, testing his knowledge, starting to publish papers in journals on all aspects of natural history from geology to botany to ornithology. There are drawings from this period MacGillivray made of microscopic fossil sections which are beautiful in their precision. He becomes fluent in conchology, in mineralogy and lichenology (in a lovely phrase in his book *A Natural History of Deeside* he wrote that he was *especially addicted to Lichenology*). He meets Darwin (then a student at Edinburgh University); to Darwin, MacGillivray appears a bit uncouth, a bit ragged-looking: *He had not much the appearance and manners of the gentleman* ... But the two absorb themselves in natural history talk and Darwin, in his autobiography, remembers MacGillivray as being *very kind to me* and giving him some rare shells for his own collection.

In 1826 there is another lurch, another breakout from the city. Perhaps the museum and/or Jameson had become too stifling, perhaps the claustrophobia of the museum grated at him. Whatever the reason, MacGillivray quits working for Jameson, flits the city to continue his *observations in the fields* ... And for a few years he disappears, lost to the hills. Income, often precarious for MacGillivray, is perhaps most insecure during this period; by 1829 he has five children and he is trying to support both them and his fieldwork through what he calls *my labour in the closet*, editing, translating, piecemeal writing.

Eighteen thirty is an important year for MacGillivray. He is settled back in Edinburgh and books and papers start to come in a flurry. His writing begins to take on the pace of his walking. In 1830 he publishes a revised edition of Withering's *A Systematic Arrangement of British Plants*, which he abridges from four volumes into one. *Lives of the Eminent Zoologists, From Aristotle to Linnaeus* quickly follows this. Then a translation from the French of *Elements of Botany and Vegetable Physiology* in 1831. In September 1830 he is appointed to the role of Conservator of the Museum of the Edinburgh Royal College of Surgeons. The following month, October 1830, John James Audubon knocks on his door to ask for MacGillivray's help in writing his *Ornithological Biography*.

The 1830s are a giddy blur of work for MacGillivray: by the end of the decade he will have published thirteen books. How he manages to juggle everything seems extraordinary. His work for Audubon, and especially his work at the Museum of the Royal College of Surgeons, is hugely demanding. At the same time there is his teaching, editing, journal publications, fieldwork, his painting and draughtsmanship. And despite all of these commitments he somehow manages to pull together what he considers his proper work. *Descriptions of the Rapacious Birds of Great Britain* is published in 1836. It is the warm-up act for what follows, his monumental five-volume *A*

History of British Birds. Volumes I–III are published between 1837 and 1840. The final two volumes are published in 1852, the same year MacGillivray died.

The breakthrough in my search for sparrowhawks came in late summer when I found a feather. For several weeks I had been coming across the aftermath of hawk kills, the downy snow of pigeon feathers sprayed across a path or snagged in a hedge. On one occasion a goldfinch, tiny black and yellow feathers in a bright wet pile at the edge of a field. There was a pattern to these kills. The hawks were hunting in a radius which took in a corner of the wood, the field and the back gardens along my street. So I slowed my wanderings down to concentrate on this area. Then one evening, walking along a track, I found a scattered bloom of pigeon feathers; not that many, but enough to suggest hawk work. I wondered if the attack had been aborted, if the sparrowhawk had swiped but failed to grip the pigeon. The way the pigeon's feathers were smeared across the track suggested this.

Many sparrowhawk attacks are aborted, the majority fail. Juvenile hawks, especially, can misjudge prey and attempt to tackle birds that are too large. A woodpigeon is too heavy for a sparrowhawk to carry far and only female hawks are large enough to tackle such heavy prey. Female sparrowhawks are substantially larger than males, the size difference between the sexes greater than in any other British raptor (males weigh around 150 g, females around 290 g). Such is the difference in weight that female sparrowhawks are known to occasionally kill and eat males. Sometimes a female sparrowhawk can lump a pigeon a short distance, but she would not be able to fly more than a few feet above the ground with such a load. So usually pigeons are plucked and eaten close to where they are killed, the tell-tale sign a circle of feathers spread around where the hawk has worked. Sometimes a large prey item such as a pigeon will not be killed outright and the hawk will simply pin it down and begin to eat while the pigeon is still alive. There are several accounts of sparrowhawks being disturbed at their plucking and their prey then flying, stumbling away.

In amongst the pigeon feathers on the track was a sparrowhawk's feather, a primary, with a deep notch and narrow leading edge. Pale white at the base along the trailing edge, brown towards the top. The reverse side much paler than the front. Five brown bands running across – interrupting – the feather's white, the hint of a sixth at the top but too faint to make out against the darker brown. I wondered if the hawk had lost the feather in the maelstrom of the attack. I picked the feather up and took it home with me.

Finding the feather was the culmination of feeling that I was always a step behind the hawks, always arriving just after they had left. So I decided, as I had done with the sea eagles on Oronsay, to stop my wanderings and instead sit tight and wait for the sparrowhawks to come to me. Twice a week, just before first light, I would position myself at the end of the track where I had found the feather, tuck myself in to a tall hedge, and wait. From there I had a clear view of the eastern and southern edges of the wood and also a large swathe of the canopy where the wood stretched away from me to the north and west.

In 1841 MacGillivray is upgraded. It is about time. By the end of the 1830s he has written himself tired. His great friend Audubon has returned to America for the last time. Money is still a struggle and by 1841 there are nine children (two have died in infancy, William and Marion's tenth and last child is born in 1842). The first three volumes of *A History of British Birds*, pummelled by the critics, are a flop and his publishers are stalling on bringing out the final two volumes. There are hints of the impact of these pressures expressed in a letter MacGillivray wrote to Audubon in 1836:

> You desire to know how I am 'going on with the world.' The world and I are not exactly as good friends as you and I, and I am not particularly desirous of being on familiar terms with it. I have got rather into difficulties this year, but I do not exactly know the state of my affairs, and must take a few days among the hills by myself before I can understand how I am situated.

The godsend, in 1841, is that MacGillivray is appointed to the Regius Chair of Civil and Natural History at Marischal College in Aberdeen. The appointment is a hugely impressive coup for MacGillivray. Competition for the post is fierce but MacGillivray gets the nod and his family move from Edinburgh to Aberdeen, to the city of his birth.

Aberdeen is: MacGillivray walking everywhere, walking his students out of the lecture hall and into the fields and hills. Now in his late forties, he teaches on the move. Students who were taught by MacGillivray recall always struggling to keep pace with him on those outdoor lectures. His advice to one student on an excursion: *Keep your knees bent as you climb a mountain. You thus avoid having to raise your body at each step.* If you passed MacGillivray walking along the street in Aberdeen between his home and college, sometimes to St Machar's Cathedral, where he loved to sit and listen to the scriptures being read, you would notice a small, thin man walking quickly. Lately he has been walking so much and sleeping so little his clothes have become loose about him. Eyes fixed on the ground in front, he would not look up as you passed one another. Usually alone, to bump into him by mistake would be to nudge the greatest ornithologist in Europe into polite apologies and earnest concern for you. He was deeply (unusually so for the time) concerned for his students. One of them wrote of MacGillivray's teaching:

> His interest in the habits of his students was remarkable. If he saw a good student careless he would remonstrate with him privately; while earnest attention gained his favour. With his rapid power of observation he could detect even a temporary lapse from diligence. His lectures were carefully written

> out, and he dictated an epitome of them once a week. Now and then he gave out a subject for an essay, say 'The Sparrow,' and he indicated a preference for a paper bearing on its habits and life on the street and on the wing. As an examiner he was patient, tender and gentle, unwilling to say an angry word. He would rather help out the hesitating student; but it was easy to see that carelessness was an abomination to him.

His lectures are so popular students and members of staff from different faculties enrol to hear them. His pupils come to him, as if he were some sort of oracle, with specimens of rock and mollusc for MacGillivray to identify for them.

In Aberdeen MacGillivray's writing is re-energised and books and papers start to flow again. *A Manual of British Ornithology* in 1842 (revised and published in a new edition in 1846); *A History of the Molluscous Animals of the Counties of Aberdeen, Kincardine and Banff* in 1843; a revised edition of Thomas Brown's *The Conchologist's Textbook* in 1845; *Portraits of Domestic Cattle of the Principal Breeds Reared in Great Britain and Ireland* in the same year. In the summer of 1850, aged fifty-four, MacGillivray makes a six-week field trip to the valley of the Dee and the Cairngorms to study the geology and botany of the region. His son Paul, and latterly one of his daughters, Isabella, accompany him and together they climb the mountains, collecting plants from the high corries.

A year after the trip to the Cairngorms MacGillivray's doctor orders him to spend the winter of 1851–2 in Torquay, rather than expose his fragile health to the blasts of an Aberdeen winter. From Torquay, he writes in the preface to volume IV of *A History of British Birds*, finally published in early spring 1852:

> As the wounded bird seeks some quiet retreat where, freed from the persecutions of the pitiless fowler, it may pass the time of its anguish in forgetfulness of the outer world, so have I, assailed by disease, betaken myself to a sheltered nook, where, unannoyed by the piercing blasts of the North Sea, I had been led to hope that my life might be protracted beyond the most dangerous season of the year. It is thus that I issue from Devonshire, the present volume which, however contains no observations of mine made there, the scenes of my labours being in distant parts of the country.

While he is still in Devon, MacGillivray's wife, Marion, dies suddenly in Aberdeen in February 1852, aged forty-seven. Audubon had died a year earlier in America. MacGillivray's health deteriorates further and he returns home to see out his final days in Aberdeen. Volume V of *A History of British Birds* is published in the summer of 1852; in its conclusion MacGillivray writes:

Commenced in hope, and carried on with zeal, though ended with sorrow and sickness, I can look upon my work without much regard to the opinions which contemporary writers may form of it, assured that what is useful within it will not be forgotten, and knowing that already it has had a beneficial effect on many of the present, and will more powerfully influence the next generation of our home ornithologists. I had been led to think that I had occasionally been somewhat rude, or at least blunt, in my criticisms; but I do not perceive wherein I have much erred in that respect, and I feel no inclination to apologise. I have been honest and sincere in my endeavours to promote the truth. With death, apparently not distant, before my eyes, I am pleased to think that I have not countenanced error, through fear of favour. Neither have I in any case modified my sentiments so as to endeavour thereby to conceal or palliate my faults. Though I might have accomplished more, I am thankful for having been permitted to add very considerably to the knowledge previously obtained of a very pleasant subject. If I have not very frequently indulged in reflections of the power, wisdom and goodness of God, as suggested by even my imperfect understanding of his wonderful works, it is not because I have not ever been sensible of the relationship between the Creator and his creatures, nor because my chief enjoyment when wandering among the hills and valleys, exploring the rugged shore of the ocean, or searching

> the cultivated fields, has not been in a sense of His presence. 'To Him who alone doeth great wonders,' be all glory and praise. Reader, farewell.

William MacGillivray dies in Aberdeen on 8 September 1852; he is fifty-six.

MacGillivray's writing, for me, is a *way* of seeing. He misses nothing. And he misses nothing out in his descriptions of the birds. His accuracy is so unflinching I use his guides to the birds to check and identify what I see (or think I've seen) in the field myself. He brings the birds into view, his descriptions more acute than any illustration or photograph. Sometimes I just want to hand everything – how to find the birds, identify them, how to write about them – over to him.

If, sometimes, he appears daunting and off-putting, to an extent he brings this on himself. In the preface to volume V he describes his ambition for *A History of British Birds*:

> Accordingly, each of our many ornithologists, real and pretended, has a method of his own, one confining himself to short technical descriptions as most useful to students, another detailing more especially the habits of the birds, as more amusing to general readers, a third viewing them in relation to human feeling and passions, a fourth converting science into romance, and giving no key to the discrimination of the species, bringing his little

> knowledge of the phenomena under the dominion of imagination, and copiously intermingling his patch-work of truth and error with scraps of poetry. The plan of this work is very different from that of any of these and is not by any means calculated to amuse the reader who desires nothing more than pleasant anecdotes, or fanciful combinations, or him who merely wishes to know a species by name. It contains the only full and detailed technical descriptions hitherto given in this country. The habits of the species are treated of with equal extension in every case where I have been enabled to study them advantageously. The internal structure has been explained in so far as I have thought it expedient to endeavour to bring it into view, and, in particular, the alimentary organs, as determining and illustrating the habits, have been carefully attended to. If imagination has sometimes been permitted to interfere, it has only been in disposing ascertained facts so as to present an agreeable picture, or to render them easily intelligible by placing them in relation to each other.

Resolute, ambitious, impatient of others who did not share his approach (some just found him rude), he did not exactly court the favour of his fellow ornithologists. So it's not surprising that they, in turn, relished the opportunity to lay into MacGillivray when *A History of British Birds* was published. But his writing – his approach – is never as austerely technical as his preface to volume V implies. And although MacGillivray may not approve of *converting science into romance*, he himself turns science into poetry on every page. Describing a night-time walk returning from Loch Muick, *a moderate weight of granite specimens* rattling in his pockets, MacGillivray reached up towards the clear sky, and wrote:

> One beautiful cluster of stars I put into my vasculum among the plants.

At a ceremony held on Tuesday 20 November, 1900 in the Natural History classroom at Marischal College, Aberdeen, several people who had been taught by him and some who had latterly been inspired by his work assembled to pay tribute to William MacGillivray and to erect a tablet in his memory. Many subscribers, among them former students and relatives of MacGillivray, contributed to the cost of the memorial tablet, and at the same time they commissioned the granite tombstone which was erected above MacGillivray's grave in the New Calton cemetery in Edinburgh. For forty-eight years, before the monument was commissioned, MacGillivray's grave had simply been marked with the letters W. M. on the grave's lower cornerstone. So the ceremony at Aberdeen, and the commissioning of the gravestone in Edinburgh, was an attempt to secure William MacGillivray's legacy, to halt the slide into oblivion. One of those who spoke at the Aberdeen ceremony, Professor Trail, who had studied MacGillivray's work whilst a student at Aberdeen University, addressed the gathering:

> I did not know Professor MacGillivray personally, but I have learned to know him in a way that, I think, perhaps not very many know him, through his works; and through these I have learned to revere the man and to love his memory ...

The last to speak was Professor J. Arthur Thomson, then the Regius Professor of Natural History at Aberdeen University:

> Every one in Britain who cares much about birds does, in a real sense, know MacGillivray, for he left a lasting mark upon ornithology. May I explain in a minute why one says so. It is because, until 1837, no one in Britain had seriously tried to found a classification, or natural system of birds except upon external characters; while MacGillivray – a trained anatomist – got far beneath the surface and showed that a bird is not always, nor altogether, to be known by its feathers ...

Add to these tributes: Alfred Newton's comments in his book *The Dictionary of Birds*:

> I may perhaps be excused for repeating my opinion that, after Willughby, Macgillivray was the greatest and most original ornithological genius save one

> (who did not live long enough to make his powers known) that this island has produced.

One night MacGillivray dreamt that four hooting owls dropped down his chimney, alighted on his dissecting table and began to rummage about in the midst of all his books and papers:

> They [*the owls*] had probably been attracted by the odour emanating from a Buzzard's skull which I had recently dissected.

In his dream the owls proceed to bicker and complain about what they find on MacGillivray's desk. They criticise his work as being too preoccupied by technicalities, as not inhabiting – not sufficiently imagining – their lives:

> 'Nothing here but dry sapless stuff. MacGillivray's Raptors, etc' observed one of the owls; 'Gutts and gizzards' quoth another, 'fit only for Turkey vultures' tedious technicalities and objectless digressions,' shrieked the third. 'Besides' said the fourth, 'the fellow ought to imitate us, he has no respect to the majesty of nature …'

The dream ends abruptly when, as MacGillivray recalls,

> Hardly knowing to laugh or cry, I awoke.

The owls are far too harsh on MacGillivray, he is anything but the tedious technician they dismiss him as. His genius, as Professor Thomson stated in his memorial address, was to get *far beneath the surface* of the birds. And it is not solely through his anatomical scalpel that he achieves this. Much more so, it is the way MacGillivray responds to the birds, the way he describes them in their natural setting, the ways he uses language to get *far beneath* the birds and lift them into being:

> A flight of sandpipers is a beautiful sight; there they wheel around the distant point, and advance over the margin of the water; swiftly and silently they glide along; now, all inclining their bodies to one side, present to view their undersurface, glistening in the sunshine; again, bending to the other side, they have changed their colour to dusky grey; a shot is fired, and they plunge with an abrupt turn, curve aside, ascend with a gliding flight, and all, uttering shrill cries, fly over the stream to settle on the shore that settles out towards Barnbogle ruins.

The great loss is that some of MacGillivray's writing has not survived. There are three journals in existence: one that records the year he spent in the Outer Hebrides between 1817 and 1818; one that records his walk from Aberdeen to London in 1819; and another which records a journey he undertook while working for the Museum of the Royal College of Surgeons, who commissioned him to visit and report on other museums in Scotland, England and Ireland in 1833.

Though only three survive, everything points to MacGillivray having kept a journal throughout his life. He spent his whole life walking, exploring the natural world; it seems improbable that he did not keep a written record of these expeditions. In his book *The Natural History of Deeside* he describes an incident where he is suddenly enveloped in a thick mist on the summit of Ben Macdui. In order to describe the mountain in more detail, MacGillivray then quotes in the book from the notes he made in his *1830* journal in which he had sketched an outline of the geology and appearance of Ben Macdui:

> Now this Ben-na-muic-dhui on which we now stand 'consists,' as I find recorded in my journal of 1830, 'of a huge rounded mass of granite, which on the western side, towards the summit, presents a corry formed by a semicircular range of precipices, the rocks of which are marked by nearly perpendicular fissures, with transverse rents, covered toward the base by débris, and sloping into a small lake named Lochan-Uaine (green little lake), the waters of which are singularly clear, and have a bluish-green tint, which has a remarkable effect as contrasted with the ordinary tints of the Scottish lakes. On these precipices, as well as on other parts of the mountain, patches of snow remain unmelted during the summer and autumn. On the opposite side the mountain declines irregularly toward the head of

> Loch Aain, terminating in a magnificent range of precipices.'

So there is proof that he kept a journal in 1830. But I'm sure there must have been shelves of his journals stretching right back to 1817 (the year he began his Hebridean journal), perhaps earlier. And I'm sure, too, that MacGillivray must have drawn on his backlist of journals as he might an encyclopaedia, to shake his memory, to check the name of a plant he had momentarily forgotten.

In the Hebridean journal of 1817–18 MacGillivray mentions a long walk he plans to make once he leaves Harris, walking through Skye, the Western Highlands and the far north of Scotland. If he did manage to complete that walk it is a great shame his journal of it does not survive. I would have loved the chance to read his account of these landscapes, their extraordinary geology, their birds and plants. To have MacGillivray as a guide to these unique, dramatic landscapes: what a thing that would have been.

The assumption is that after MacGillivray's death his son, Paul, took all of his father's journals and papers with him when he emigrated to Australia and that the journals were later destroyed in a fire there. If this is indeed what happened to MacGillivray's journals then it is a tremendous loss to the natural history writing of these islands. But read them – those journals that have survived – and, like Professor Trail in his memorial address, you cannot help but *to revere the man and to love his memory*.

> **Sparrowhawk**
> There it comes, silently and swiftly gliding, at the height of a few feet, over the grass field, now shooting along the hedge, now gliding over it to scan the other side, and again advancing with easy strokes of its half-expanded wings. A beautiful machine it is certainly, and marvellously put together, to be nothing but a fortuitous concourse of particles, as some wise men, believing no such thing themselves, would have us to believe. As if suspecting the concealment of something among the grass, it now hovers a while, balancing itself with rapid but gentle beats of its wings, and a vibratory motion of its expanded tail; but, unable to discover any desirable object, away it speeds, bounds over the stone wall, and curving upwards alights on that stunted and solitary ash, where it stands in a nearly erect posture, and surveys the neighbourhood.

From now on, this is my routine. I slip into the hedge at the end of the track while it is still dark and watch the light rinse the field. Who is here? First up, a roe deer and her well-grown calf. They are wading back through the high dewy grass towards the wood. The calf is skittish, all sprints and feints around its mother. It is September and their pelts are a deep red, turmeric and umber colour. Next come the kestrels, a pair, flying low across the field, chittering, yattering, chasing each other. Then pigeon after pigeon, breaking from the wood at speed. Each sudden burst of pigeon briefly holds the possibility of a hawk hidden in its shape. But it never is, and when it comes the hawk is altogether something else. Different shape, different poise, different bent: hawk, it could not be anything else. The sparrowhawk lands on the dead branch of an ash tree on the edge of the wood, quivers there. Straight away, like a conjuror's hat trick, another hawk is pulled out of the wood and lands beneath the first. When it lands it tips the other hawk from its perch and then the pair of them are off, skittering, chasing after each other through the trees.

The next half-hour is a whirl of hawks, flickering, twisting, speeding through the trees around the edge of the wood, careering through the branches. Glimpses of them: racing, swerving through the foliage, dark long-winged shapes. Here they come again, one after the other, popping out of the wood to perch together in the ash tree. But too restless to stay there for long, they are off again skidding through the trees. They move so quickly it's hard to get the colours of their plumage right but the impression is of a mostly brown dorsal and a pale breast. I also find it hard to judge if they are a pair, a male and female. I presume they are and that I have just caught them emerging from their roost in the wood. But, again, the speed at which they chase after each other makes it hard to appreciate the size difference between the sexes. I notice, when the birds are perched, how they blend in amongst the brown clumps of dead ash key seeds, I keep mistaking the seeds for hawks.

A kestrel comes in. He is much stiffer in flight than the hawks, sharper, narrower-winged, nothing of the hawks' fluidity. The sharp angles of his wings make him seem a little larger than the sparrowhawks. The kestrel lands at the top of the ash tree and his tail bobs up behind when he lands, as if the action of the tail balances his landing. Immediately the hawks join him, shuffling around on the lower branches. For a while I watch a laddering of raptors: kestrel – then hawk – then hawk below.

The sparrowhawk is not a specialist. It does not rely on a particular prey species, as the honey buzzard does with wasps or the kestrel with field voles. It is, however, an avian specialist, as its diet is almost entirely made up of small to medium-sized birds. Some small mammals are taken, young rabbits early in the season, field voles especially in plague years, bank voles in the woods. But live birds predominate. It is also, like its larger accipiter relative, the goshawk, a woodland specialist, able to pursue prey at speed through the narrow airspace of a wood. A wide variety of bird species are predated: everything from wren (rarely) to woodpigeon (often). But locally, certain species will provide the bulk of the sparrowhawk's diet. Fledgling songbirds in June and July are crucial. Sparrowhawks synchronise their nesting to coincide with the availability of songbird fledglings, and without this glut of vulnerable prey in the summer months the hawk's ability to breed successfully would be severely impaired.

Important prey species in the British Isles include starling, chaffinch, song thrush, blackbird and woodpigeon, especially in spring and summer; fieldfare and redwing in winter. Birds of deep cover like the wren and blue tit are far less vulnerable to sparrowhawk predation than more conspicuous species of open spaces like the robin and redwing. Vocal, displaying songbirds in springtime are especially vulnerable. Male hawks take a higher proportion of smaller prey species than the females. The males also tend to hunt in more concentrated woodland; female hawks tend to range further afield and hunt in more open landscapes.

The sparrowhawk is not the efficient predator of garden birds we tend to assume it is; its prey species are efficient at avoiding it. Alarm calls warn of a hawk's approach and give the chance for smaller birds to seek cover – about three seconds is all they need to make their escape. With the slightest head-start most birds will escape, as the hawk cannot follow them into dense foliage. A woodpigeon can outfly a sparrowhawk with relative ease. Once its prey has put sufficient distance between itself and the hawk, the sparrowhawk does not waste its energy and, unlike a merlin, usually gives up the pursuit. The hawk's agility does not always mean it can outmanoeuvre smaller, more agile prey. Starvation is a major cause of mortality for sparrowhawks, especially between March and April, the hungry gap, when winter migrants have left these islands and summer migrants have not yet arrived.

Sparrowhawks can be a major cause of mortality for certain birds during certain narrow windows in the year. These windows are defined by the stage of abundance and vulnerability of the prey. So fledglings, when they have just left the nest, are at their most vulnerable to predation. Despite this heavy predation, as Ian Newton explains in his great study of the raptor, there is no evidence that sparrowhawks have any long-term effect on the size of their prey's population. In other words, if the hawks were removed from the environment, there would be no noticeable increase in their prey species. In the absence of their major predator, other mortality or 'controlling' factors, starvation in particular, would come into play on these birds. The best evidence for this occurred when sparrowhawks *were* removed from large parts of the British countryside in the late 1950s and early 1960s through poisoning by organochlorine pesticides. During this period, despite the absence of their major predator, the songbird population remained the same. Neither, when sparrowhawks returned to the ecosystem, was there any noticeable decline in their prey species.

To catch its prey the sparrowhawk relies on stealth and surprise. It uses obstacles, hedges and walls for instance, to conceal itself, at the last minute rushing up over a hedge to attack prey on the other side. It flies as close to the ground as possible and conceals its approach by flying, when conditions allow, with the sun behind it. I have seen a sparrowhawk fly so low up a road in front of me there seemed to be no visible gap between bird and tarmac.

The reason you rarely see sparrowhawks is because to hunt – to survive – they rely on concealment. Sparrowhawks can (more rarely in Britain than on mainland Europe) suffer predation themselves, so concealment also helps to negate this risk (goshawks, peregrines and tawny owls have all been known to kill sparrowhawks). If a hawk perched on the top of a tree like a kestrel does, every songbird in the neighbourhood would yell: *Hawk!* The sparrowhawk must always be just out of sight; that is where you will find them.

I am in the hedge before dawn. I'm glad of the track, the way it allows me to avoid the wood, to slip into the hedge without waking the wood. I'm just in time to catch the change of shift: a muntjac crossing the field, heading back to the wood; crows and jackdaws, coming the other way, pour from the trees like black smoke.

The early morning soundscape (in this order): tawny owl, the screech-echo of a pair; jackdaw chatter; a moorhen's alarm from the moat behind the track; a raven's croak; pheasant (very loud inside the wood) sets off another pheasant a whole field away; a buzzard calling on the soar; a magpie's rattle ... Next comes the staggered departure from the roost: crows and jackdaws the first to leave in their tatterdemalion flocks, magpies in their slipstream, then the kestrel, keeking, wide awake.

The dawn itself is split between the buzzard-light, which is a murky pre-dawn light, and the hawk-light, which comes soon after dawn. The buzzard is always out of the roost before the hawk, hunting when there seems barely enough light to see in. It is only when the light has really broken through, cleaned and sharpened the edges of the wood, that the sparrowhawk comes out of the trees from its roost, silent, quick, broad-winged. Sometimes a single hawk, sometimes a pair, chasing each other low across the field, turning knots through one another, playful, sinuous like otters. Usually just a fleeting glimpse: the hawk bursts out of the trees, shoots low across the field and is gone.

But that brief glimpse is enough. I long for it, and long to get back to the hedge. Some mornings I do not see the hawks, but because they use the wood to roost, I often catch the birds when they depart at dawn. And so, over the weeks and months, I build up a store of hawk sightings, a steady accumulation of wonder. One morning I return to the hedge to find it has been cut right back by a tractor's hedge cutter: it is almost unrecognisable, a shredded, splintered mess. I feel miserable and a little silly that I have become so possessive of a hedge. But I realise it will grow back and thicken up in the spring and there is still space for me to slot in and conceal my shape.

There are some mornings when the wind seems to get inside the birds. Jackdaws are bundled out of the wood in erratic, swirling columns. Pigeons pour out of the trees in their hundreds. A red kite comes over high and fast through the dawn with the wind behind it. A tiny goldcrest is made larger – rounder – by the breeze puffing out its feathers.

A hawk goes up from the wood, squabbles with a crow, then dives fast and steep back into the centre of the wood. Then she is climbing again, spinning up from the trees. She banks slowly around above the wood and out across the field. Another hawk joins her there and they reel together high above the village.

Because I am so absorbed in watching the hawks, I do not notice that a deer has walked right up to me. So when I look down, I am astonished to see her: a diminutive muntjac, just three feet away, her black eyes staring at me. I freeze, but the deer hasn't got a hold of me, hasn't seen me properly. The hedge, despite its crew cut, seems to still disguise me. The muntjac ambles past; stiff, stumpy gait, lumpy brown coat. Then my scent pricks her: she kicks her hind feet up, and bucks away across the field, white rump flashing as she goes. When she reaches the wood she slows to a trot, glances back, barks, then hurries on into the wood. For a while I can see her white tail signalling, like a light fading inside the trees.

I leave the hedge and walk back along the muddy track. Blackbirds are out along the path, scraping through the leaf litter. A moorhen makes a dash for cover. Through the gate, turn right, and up the street. Coat off, kettle on, binoculars back in their case.

Upstairs, on my desk, I keep a jar of feathers – kestrel, magpie, buzzard, pheasant – keepsakes from the fields and wood. My children also bring me feathers they have found, knowing that even a bedraggled pigeon feather fished from a drain will make me smile. The collection keeps growing; sometimes, when the light inside the room is thin and closing, it looks as if my desk has grown a wing.

Through the window three magpies have lit up the pale winter lawn. In the branches of the apple tree – the old, gnarled Russet – a clump of mistletoe is a ball of green light. There is a faint hum inside the room; the fan on my laptop, which I barely notice, has noticed one of the feathers – a buzzard's – and is breathing gently against it. The downy barbs around the feather's quill shiver and lift.

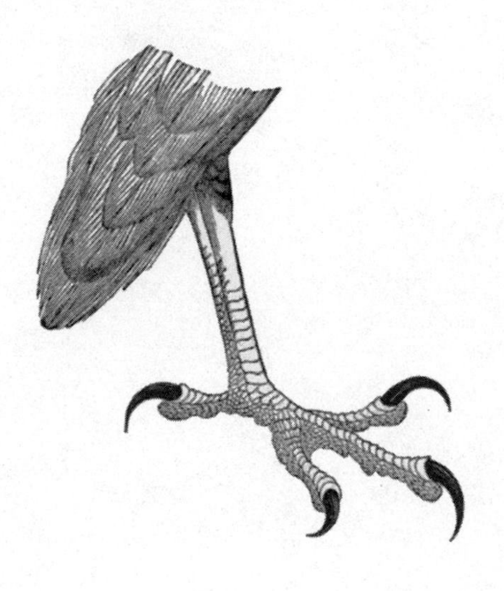

Bibliography

I Hen Harrier

Armstrong, Edward A. – *The Folklore of Birds* (Dover, 1970)

Balfour, Edward – 'Observations on the Breeding Biology of the Hen Harrier in Orkney' (*Bird Notes*, 27, no. 6, 177–183, no. 7, 216–224, 1957)

Barkham, Patrick – 'The mystery of the missing hen harriers' (*Guardian*, 13 January 2015)

Berry, R. J. – *Orkney Nature* (T. & A. D. Poyser, 2000)

Bevis, John – *Direct from Nature: The Photographic Work of Richard and Cherry Kearton* (Colin Sackett, 2007)

Cuthbert, Olaf – *Eddie: An Orkney Ornithologist Remembered* (Felix Books, 2005)

Eagle, Raymond – *Seton Gordon: The Life and Times of a Highland Gentleman* (Lochar, 1991)

Fenton, Alexander – *The Northern Isles: Orkney and Shetland* (John Donald, 1978)

Firth, John – *Reminiscences of an Orkney Parish* (W. R. Rendall, 1974)

Gordon, Seton – *Thirty Years of Nature Photography* (Cassell, 1936)

— 'Bird Photography 70 Years Ago' (*Country Life* Magazine, 8 April 1976)

Hamerstrom, Frances – *Harrier, Hawk of the Marshes: The Hawk that is Ruled by a Mouse* (Smithsonian Institution Press, 1986)

Hedges, John W. – *Isbister: A Chambered Tomb in Orkney* (B.A.R. 115, 1983)

— *Tomb of the Eagles: A Window on Stone Age Tribal Britain* (John Murray, 1987)

Kearton, Richard – *With Nature and a Camera* (Cassell, 1902)

Mackay Brown, George – *An Orkney Tapestry* (Quartet Books, 1978)

Meek, E. R., Rebecca, G. W., Ribbands, B. and Fairclough, K. – 'Orkney Hen Harriers: a major population decline in the absence of persecution' (*Scottish Birds*, 19, no. 5, 290–298, 1998)

Morrow, Phyllis and Volkman, Toby – 'The Loon with the Ivory Eyes: A Study in Symbolic Archaeology' (*Journal of American Folklore*, 88, no. 348, 143–150, 1975)
Pitts, M. – 'Flight of the Eagles' (*British Archaeology*, 86, 6, January–February 2006)
Rainey, Froelich – 'The Ipiutak Culture at Point Hope, Alaska' (*American Anthropologist*, 43, no. 3, 364–375, 1941)
Redpath, S. M. and Thirgood, S. J. – *Birds of Prey and Red Grouse* (Stationery Office, 1997)
Scott, Don – *The Hen Harrier: In the Shadow of Slemish* (Whittles, 2010)
Scottish Natural Heritage – *Substitute Feeding of Hen Harriers on Grouse Moors: A Practical Guide* (SNH, 1999)
Scrope, William – 'Days and Nights of Salmon-Fishing in the Tweed, with a short Account of the Natural History and Habits of the Salmon, Instructions to Sportsmen, Anecdotes, &c' (*Quarterly Review*, 77, December 1845)
Simmons, Robert E. – *Harriers of the World: Their Behaviour and Ecology* (Oxford University Press, 2000)
Taylor, Timothy – *The Buried Soul: How Humans Invented Death* (Fourth Estate, 2002)
Thomson, David – *The People of the Sea: A Journey in Search of the Seal Legend* (Barrie & Rockliff, 1965)
Watson, Donald – *The Hen Harrier* (T. & A. D. Poyser, 1977)
— *In Search of Harriers* (Langford, 2010)
Whyte, Craig – *The Wildlife of Rousay, Egilsay and Wyre* (Brinnoven, 2004)
Wilkinson, Benjamin Joel – *Carrion Dreams 2.0: A Chronicle of the Human–Vulture Relationship* (Abominationalist Productions, 2012)
Willis, Douglas P. – *Moorland and Shore: Their Place in the Human Geography of Old Orkney* (Department of Geography, University of Aberdeen, 1983)

II Merlin

Avery, Mark and Leslie, Roderick – *Birds and Forestry* (T. & A. D. Poyser, 1990)
Bibby, C. J. and Nattrass, M. – 'Breeding status of the Merlin in Britain' (*British Birds*, 79, no. 4, 170–185, 1986)
Crampton, C. B. – *The Vegetation of Caithness Considered in Relation to the Geology* (Committee for the Survey and Study of British Vegetation, 1911)
Land, Michael F. and Nilsson, Dan-Eric – *Animal Eyes* (Oxford University Press, 2002)
Lindsay, R. A., Charman, D. J., Everingham, F., O'Reilly, R. M., Palmer, M. A., Rowell, T. A., Stroud, D. A. (ed. Ratcliffe, D. A. and Oswald, P. H.) – *The Flow Country: The Peatlands of Caithness and Sutherland* (Nature Conservancy Council, 1988)
Mabey, Richard – *Home Country* (Century, 1990, pp. 119–131, 'Away Games')

Munro, Henrietta – *They Lived by the Sea: Folklore and Ganseys of the Pentland Firth* (Henrietta Munro and Rae Compton, 1983)

Nethersole-Thompson, Desmond and Maimie – *Greenshanks* (Buteo Books, 1979)

Newton, I., Meek, E. R. and Little, B. – 'Breeding Ecology of the Merlin in Northumberland' (*British Birds*, 71, no. 9, 376–398, 1978)

Nicolaisen, W. F. H. – *Scottish Place-Names* (John Donald, 2001)

Orchel, Jack – *Forest Merlins in Scotland: Their Requirements and Management* (The Hawk and Owl Trust, 1992)

Orton, D. A. – *The Merlins of the Welsh Marches* (David & Charles, 1980)

Proctor, Noble S. and Lynch, Patrick J. – *Manual of Ornithology: Avian Structure and Function* (Yale University Press, 1993)

Ross, David – *Scottish Place-names* (Birlinn, 2001)

Rowan, W. – 'Observations on the Breeding-Habits of the Merlin' (*British Birds*, 15, no. 6, 122–129, no. 9, 194–202, no. 10, 222–231, no. 11, 246–253, 1921–22)

Sale, Richard – *The Merlin* (Snowfinch, 2015)

Stroud, David A., Reed, T. M., Pienkowski, M. W., Lindsay, R. A. (ed. Ratcliffe, D. A. and Oswald, P. H.) – *Birds, Bogs and Forestry: The Peatlands of Caithness and Sutherland* (Nature Conservancy Council, 1988)

Trobe, W. M. – *The Merlin* (Oriel Stringer, 1990)

Withers, Charles W. J. – *Gaelic in Scotland 1698–1981: The Geographical History of a Language* (John Donald, 1984)

Wright, Peter M. – 'Distribution, site occupancy and breeding success of the Merlin (*Falco columbarius*) on Barden Moor and Fell, North Yorkshire' (*Bird Study*, 44, no. 2, 182–193, 1997)

— *Merlins of the South-East Yorkshire Dales* (Tarnmoor Publications, 2005)

III Golden Eagle

Campbell, Laurie and Dennis, Roy – *Golden Eagles* (Colin Baxter Photography, 1996)

Gordon, Seton – *Days with the Golden Eagle* (Williams & Norgate, 1927)

— *The Golden Eagle: King of Birds* (Collins, 1955)

Harvie-Brown, J. A. and Buckley, T. E. – *A Vertebrate Fauna of the Outer Hebrides* (David Douglas, 1888)

Haworth, Paul F., Mcgrady, Michael J., Whitfield, D. Philip, Fielding, Alan H. and McLeod, David R. A. – 'Ranging distance of resident Golden Eagles (*Aquila chrysaetos*) in western Scotland according to season and breeding status' (*Bird Study*, 53, no. 3, 265–273, 2006)

Hunter, James – *The Making of the Crofting Community* (John Donald, 1982)

Lawson, Bill – *The Teampull at Northton and The Church at Scarista: Harris Churches in their Historical Setting* (Bill Lawson Publications, 1993)

Lockie, J. D. – 'The Breeding Density of the Golden Eagle and Fox in Relation to Food Supply in Wester Ross, Scotland' (*Scottish Naturalist*, 71, no. 2, 67–77, 1964)
— and Stephen, D. – 'Eagles, Lambs and Land Management on Lewis' (*Journal of Animal Ecology*, 28, no. 1, 43–50, 1959)
Love, John A. – *Eagles* (Whittet Books, 1989)
Lynch, Michael – *Scotland: A New History* (Pimlico, 1995)
MacGillivray, William – *A Hebridean Naturalist's Journal 1817–1818* (Acair, 1996)
Macpherson, H. B. – *The Home Life of a Golden Eagle* (Witherby & Co., 1909)
Moisley, H. A. – *Uig: A Hebridean Parish* (Geographical Field Group, 1961)
Newton, Ian – *Population Ecology of Raptors* (T. & A. D. Poyser, 1979)
Tomkies, Mike – *On Wing and Wild Water* (Jonathan Cape, 1987)
— *Golden Eagle Years* (Jonathan Cape, 1994)
Watson, Jeff – *The Golden Eagle* (Second Edition) (T. & A. D. Poyser, 2010)
Whitfield, D. P., Fielding, A. H., McLeod, D. R. A. and Haworth, P. F. – *A conservation framework for golden eagles: implications for their conservation and management in Scotland* (Scottish Natural Heritage Commissioned Report no. 193, 2008, [ROAME no. F05AC306])
Whitfield, P. – 'Golden eagle (*Aquila chrysaetos*) ecology and conservation issues' (*Scottish Natural Heritage Review*, no. 132, 2000)

IV Osprey

Abbott, C. G. – *The Home-Life of the Osprey* (Witherby & Co., 1911)
Bagnold, R. A. – *The Physics of Blown Sand and Desert Dunes* (Chapman & Hall, 1971)
Bain, George – *History of Nairnshire* (Nairnshire Telegraph, 1893)
Ball, Philip – *Flow: Nature's Patterns – A Tapestry in Three Parts* (Oxford University Press, 2009)
Brown, Philip – *The Scottish Ospreys: From Extinction to Survival* (William Heinemann, 1979)
— and Waterston, George – *The Return of the Osprey* (Collins, 1962)
Dennis, Roy – *A Life of Ospreys* (Whittles, 2008)
Fenech, Natalino – *Fatal Flight: The Maltese Obsession with Killing Birds* (Quiller Press, 1992)
Forestry Commission Scotland – *The History of Culbin* (Forestry Commission, 2012)
Gordon, Seton – *In Search of Northern Birds* (Eyre & Spottiswoode, 1941)
MacGregor, Alasdair Alpin – *The Buried Barony* (Robert Hale, 1949)
Macpherson, H. A. – *The Visitation of Pallas's Sand-Grouse to Scotland in 1888, together with an account of its nesting, habits, and migrations* (R. H. Porter, 1889)
— *A Vertebrate Fauna of Lakeland* (David Douglas, 1892)
Poole, Alan F. – *Ospreys: A Natural and Unnatural History* (Cambridge University Press, 1989)

Rubinstein, Julian – 'Operation Easter: The Hunt for Illegal Egg Collectors' (*New Yorker*, 22 July 2013)
St John, Charles – *The Wild Sports and Natural History of the Highlands* (John Murray, 1948, first published 1846)
— *A Tour in Sutherlandshire* (David Douglas, 1884)
Thomson, David – *Nairn in Darkness and Light* (Hutchinson, 1987)

V Sea Eagle

Cameron, A. D. – *Go Listen to the Crofters: The Napier Commission and Crofting a Century Ago* (Acair, 1986)
Craig, David – *On the Crofter's Trail* (Pimlico, 1997)
Crawford, Carol L. – *Bryophytes of Native Woods: A Field Guide to the Common Mosses and Liverworts of Scotland and Ireland's Native Woodlands* (Native Woodlands Discussion Group, 2002)
Drever, James – '"Taboo" Words Among Shetland Fishermen' (*Old-Lore Miscellany of Orkney, Shetland, Caithness and Sutherland*, 10, part 6, 235–240, 1946)
Edmondston, Thomas – *An Etymological Glossary of the Shetland & Orkney Dialect; with some derivations of names of places in Shetland* (Adam & Charles Black, 1866)
Fenton, Alexander – 'The Tabu Language of the Fishermen of Orkney and Shetland' (*Ethnologia Europaea*, 2–3, 118–122, 1968–69)
Fraser Darling, F. – *Natural History in the Highlands & Islands* (Collins, 1947)
Gaskell, Philip – *Morvern Transformed: A Highland Parish in the Nineteenth Century* (Cambridge University Press, 1968)
Gelling, Margaret – *Place-Names in the Landscape: The Geographical Roots of Britain's Place-Names* (Phoenix Press, 2000)
Gray, Robert – *The Birds of the West of Scotland, Including the Outer Hebrides* (Thomas Murray & Son, 1871)
Harvie-Brown, J. A. and Buckley, T. E. – *A Vertebrate Fauna of Argyll and the Inner Hebrides* (David Douglas, 1892)
Jakobsen, Jakob – *An Etymological Dictionary of the Norn Language in Shetland* (David Nutt, 1928)
Lodge, George E. – *Memoirs of an Artist Naturalist* (Gurney and Jackson, 1946)
Love, John A. – *The Return of the Sea Eagle* (Cambridge University Press, 1983)
— *A Saga of Sea Eagles* (Whittles, 2013)
McClure, J. Derrick – 'Distinctive semantic fields in the Orkney and Shetland dialects, and their use in the local literature', in Millar, Robert McColl (ed.) – *Northern Lights, Northern Words. Selected Papers from the FRLSU Conference, Kirkwall 2009* (Forum for Research on the Languages of Scotland and Ireland, 58–69, 2010)
Mackenzie, Osgood – *A Hundred Years in the Highlands* (The National Trust for Scotland, 1988)

Macleod, Norman – *Morvern: A Highland Parish* (Birlinn, 2002, first published 1863)
Marquiss, M., Madders, M., Irvine, J. and Carss, D. N. – *The Impact of White-tailed Eagles on Sheep Farming on Mull: Final Report* (Scottish Executive, 2004)
Martin, Martin – *A Description of the Western Islands of Scotland Circa 1695* (Birlinn, 1999)
Morton Boyd, John and Bowes, D. R. (eds) – *The Natural Environment of the Inner Hebrides* (The Royal Society of Edinburgh, 1983)
Murray, W. H. – *The Hebrides* (Heinemann, 1966)
Napier Commission – *Evidence Taken by Her Majesty's Commissioners of Inquiry into the Conditions of the Crofters and Cottars in the Highlands and Islands of Scotland* (Neill & Company, 1884)
— *Report of Her Majesty's Commissioners of Inquiry into the Condition of the Crofters and Cottars in the Highlands and Islands of Scotland. With Appendices* (Neill & Company, 1884)
Saxby, Henry L. – *The Birds of Shetland with Observations on their Habits, Migration, and Occasional Appearance* (Maclachlan & Stewart, 1874)
Scottish Natural Heritage – *Is lamb survival in the Scottish Uplands related to the presence of breeding white-tailed eagles (Haeliaeetus albicilla) as well as other livestock predators and environmental variables: A pilot study into sea eagle predation on lambs in the Gairloch area* (Commissioned Report no. 370, SNH, 2010)
Thornber, Iain – *The Gaelic Bards of Morvern* (Iain Thornber, 1985)
Willgohs, Johan Fr. – *The White-tailed Eagle Haliaëtus Albicilla Albicilla (Linné) in Norway* (Norwegian Universities Press, 1961)

VI Goshawk

Aitken, A. J. – 'Scots', in McArthur, Tom (ed.), *The Oxford Companion to the English Language* (Oxford University Press, 1992)
Diener, Alexander C. and Hagen, Joshua – *Borders: A Very Short Introduction* (Oxford University Press, 2012)
Forestry Commission – *Britain's Forests: Kielder* (HMSO, 1950)
Glauser, Beat – *The Scottish-English Linguistic Border* (Francke Verlag, Bern, 1974)
Jameson, Conor Mark – *Looking for the Goshawk* (Bloomsbury, 2013)
Kay, Billy – *Scots: The Mither Tongue* (Mainstream, 2006)
Kenward, Robert – *The Goshawk* (T. & A. D. Poyser, 2007)
Kenward, R. E. and Lindsay, I. M. (eds) – *Understanding the Goshawk* (The International Association for Falconry and Conservation of Birds of Prey, 1981)
Lamont, Claire and Rossington, Michael (eds) – *Romanticism's Debatable Lands* (Palgrave Macmillan, 2007)
Llamas, Carmen – 'Convergence and Divergence Across a National Border', in Llamas, Carmen and Watt, Dominic (eds), *Language and Identities* (Edinburgh University Press, 2010)

McCulloch, Christine – 'Dam Decisions and Pipe Dreams: The Political Ecology of Reservoir Schemes (Teesdale, Farndale and Kielder Water) in North East England' (D.Phil. Thesis, University of Oxford, 2005)
Macdonald, Helen – *H is for Hawk* (Jonathan Cape, 2014)
Mack, James Logan – *The Border Line: From the Solway Firth to the North Sea, along the Marches of Scotland and England* (Oliver & Boyd, 1926)
Mackay Mackenzie, W. – 'The Debateable Land' (*Scottish Historical Review*, 30, no. 110, 109–125, 1951)
Marquiss, M. and Newton, I. – 'The Goshawk in Britain' (*British Birds*, 75, no. 6, 243–260, 1982)
Neville, Cynthia J. – *Violence, Custom and Law: The Anglo-Scottish Border Lands in the Later Middle Ages* (Edinburgh University Press, 1998)
Northumbrian Water Authority – *Kielder Water Scheme: A Bibliography* (Library and Information Service, 1982)
Robson, Eric – *The Border Line* (Frances Lincoln, 2006)
White, T. H. – *The Goshawk* (Jonathan Cape, 1953)
Wilson, Keith and Leathart, Scott (eds) – *The Kielder Forests: A Forestry Commission Guide* (Forestry Commission, 1982)

VII Kestrel

Baillie, S. R., Marchant, J. H., Leech, D. I., Massimino, D., Eglington, S. M., Johnston, A., Noble, D. G., Barimore, C., Kew, A. J., Downie, I. S., Risely, K. and Robinson, R. A. – *Bird Trends 2012: Trends in numbers, breeding success and survival for UK breeding birds* (British Trust for Ornithology Research Report, 644, 2013)
Dobinson, Colin – *Fields of Deception: Britain's Bombing Decoys of World War II* (Methuen, 2000)
Evans, Darren M., Redpath, Stephen M., Elston, David A., Evans, Sharon A., Mitchell, Ruth J. and Dennis, Peter – 'To graze or not to graze? Sheep, voles, forestry and nature conservation in the British uplands' (*Journal of Applied Ecology*, 43, no. 3, 499–505, June 2006)
Hesketh, Phoebe – *My Aunt Edith* (Lancashire County Books, 1992)
Hutchinson, Roger – *The Soap Man: Lewis, Harris and Lord Leverhulme* (Birlinn, 2003)
Lane, Dave – *Winter Hill Scrapbook* (Dave Lane, 2009)
Nicolson, Nigel – *Lord of the Isles: Lord Leverhulme in the Hebrides* (Weidenfeld & Nicolson, 1960)
Rawlinson, John – *About Rivington* (Nelson Brothers, 1981)
Riddle, Gordon – *Seasons with the Kestrel* (Blandford, 1992)
Risely, K., Massimino, D., Newson, S. E., Eaton, M. A., Musgrove, A. J., Noble, D. G., Procter, D. and Baillie, S. R. – *The Breeding Bird Survey 2012* (BTO Research Report, 645, 2013)
Salveson, Paul – *Will Yo Come O Sunday Mornin? The 1896 Battle for Winter Hill* (Red Rose Publishing, 1982)
Shrubb, Michael – *The Kestrel* (Hamlyn, 1993)

Smith, M. D. – *Leverhulme's Rivington* (Nelson Brothers, 1984)
— *About Horwich* (Nelson Brothers, 1988)
Tattersall, F. H., Avundo, A. E, Manley, W. J., Hart. B. J. and Macdonald, D. W. – 'Managing set-aside for field voles (*Microtus agrestis*)' (*Biological Conservation*, 96, no. 1, 123–128, 2000)
Videler, J. J., Weihs, D. and Daan, S. – 'Intermittent gliding in the hunting flight of the kestrel (*Falco tinnunculus* L.)' (*Journal of Experimental Biology*, 102, 1–12, 1983)
Village, Andrew – *The Kestrel* (T. & A. D. Poyser, 1990)

VIII Montagu's Harrier

Arroyo, Beatriz – 'Breeding Ecology and Nest Dispersion of Montagu's Harrier (*Circus pygargus*) in Central Spain' (D.Phil. Thesis, University of Oxford, 1995)
Childers, J. W. (ed.) – *Lord Orford's Voyage round the Fens in 1774* (Cambridge Libraries Publications, 1988)
Clarke, Roger – *Harriers of the British Isles* (Shire Natural History, 1990)
— *Montagu's Harrier* (Arlequin Press, 1996)
Darby, H. C. – *The Changing Fenland* (Cambridge University Press, 1983)
Dee, Tim – *Four Fields* (Jonathan Cape, 2013)
Delamain, Jacques – *Why Birds Sing* (Victor Gollancz, 1932)
Dormer, Sally – *Twenty Objects for Twenty Years: The Ramsey Abbey Censer and Incense Boat* (Victoria and Albert Museum website, July 2013)
Fowler, Gordon – 'The Extinct Waterways of The Fens' (*Geographical Journal*, 83, no. 1, 30–39, 1934)
Garcia, J. T. and Arroyo, B. E. – 'Food-niche differentiation in sympatric Hen (*Circus cyaneus*) and Montagu's Harriers (*Circus pygargus*)' (*Ibis*, 147, no. 1, 144–154, 2005)
Godwin, Harry – *Fenland: Its Ancient Past and Uncertain Future* (Cambridge University Press, 1978)
Image, Bob – 'Montagu's Harriers taking prey disturbed by farm machinery' (*British Birds*, 85, no. 10, 559, 1992)
Kingsley, Charles – *Prose Idylls* (Macmillan, 1873)
Miller, Samuel H. and Skertchly, Sydney B. J. – *The Fenland Past and Present* (Longmans, Green & Co., 1878)
Montagu, George – *Ornithological Dictionary* (J. White, 1802)
— 'Some interesting Additions to the Natural History of *Falco cyaneus* and *pygargus*, together with Remarks on some other British Birds' (*Ornithological Papers, Transactions of the Linnean Society*, vol. IX, 1808)
Robinson, Eric and Fitter, Richard (eds) – *John Clare's Birds* (Oxford University Press, 1982)
Rosén, Mikael – *Birds in the Flow: Flight Mechanics, Wake Dynamics and Flight Performance* (Department of Ecology – Animal Ecology, Lund University, Sweden, 2003)

Simmons, Robert E. – *Harriers of the World: Their Behaviour and Ecology* (Oxford University Press, 2000)
Skertchly, Sydney – *The Geology of the Fenland* (HMSO, 1877)
Stevenson, Henry – *The Birds of Norfolk* (John Van Voorst, 1866)
Summers, Dorothy – *The Great Level: A History of Drainage and Land Reclamation in the Fens* (David & Charles, 1976)
Weis, Henning – *Life of the Harrier in Denmark* (Wheldon & Wesley, 1923)
Wells, W. – 'The Drainage of Whittlesea Mere' (*Journal of the Royal Agricultural Society of England*, 21, 134–153, 1860)

IX Peregrine Falcon

Baker, J. A. – *The Peregrine* (Collins, 1967)
Bird, David M., Varland, Daniel E., Negro, Juan Jose (eds) – *Raptors in Human Landscapes: Adaptations to Built and Cultivated Environments* (Academic Press, 1996)
Drewitt, Ed – *Urban Peregrines* (Pelagic, 2014)
Frank, Saul – *City Peregrines: A Ten-year Saga of New York City Falcons* (Hancock House, 1984)
Harkin, Trevor – *Coventry 14th/15th November 1940 Casualties, Awards and Accounts* (War Memorial Park Publications, 2010)
Howard, R. T. – *Ruined and Rebuilt: The Story of Coventry Cathedral 1939–1962* (The Council of Coventry Cathedral, 1962)
Longmate, Norman – *Air Raid: The Bombing of Coventry 1940* (Arrow Books, 1979)
Macdonald, Helen – *Falcon* (Reaktion Books, 2006)
The Provost of Coventry Cathedral – *The Story of the Destruction of Coventry Cathedral November 14th, 1940* (Edwards, 1941)
Ratcliffe, Derek – *The Peregrine Falcon* (Second Edition) (T. & A. D. Poyser, 1993)
Ratcliffe, D. A. – 'Changes Attributable to Pesticides in Egg Breakage Frequency and Eggshell Thickness in some British Birds' (*Journal of Applied Ecology*, 7, no. 1, 67–115, 1970)
Stirling-Aird, Patrick – *Peregrine Falcon* (Bloomsbury, 2015)
Treleaven, R. B. – *Peregrine: the private life of the Peregrine Falcon* (Headland, 1977)
— *In Pursuit of the Peregrine* (Tiercel Publishing, 1998)

X Red Kite

Carrell, Severin – 'Scottish bird of prey colony hit by mass poisonings' (*Guardian*, 3 April 2014)
Carter, Ian – *The Red Kite* (Arlequin Press, 2001)
Condry, William – *The Natural History of Wales* (Collins, 1981)
Cross, Tony and Davis, Peter – *The Red Kites of Wales* (Subbuteo Natural History Books, 2005)
Davis, Peter – 'The Red Kite in Wales: setting the record straight' (*British Birds*, 86, no. 7, 295–298, 1993)

Davis, P. E. and Davis, J. E. – 'The food of the Red Kite in Wales' (*Bird Study*, 28, no. 1, 33–40, 1981)
— and Newton, I. – 'Population and Breeding of Red Kites in Wales Over a 30-Year Period' (*Journal of Animal Ecology*, 50, 759–772, 1981)
Evans, A. H. (ed.) – *Turner on Birds: A Short and Succinct History of the Principal Birds Noticed by Pliny and Aristotle, First Published by Doctor William Turner, 1544* (Cambridge University Press, 1903)
Evans, John – *The Red Kite in Wales* (Christopher Davies, 1990)
Forestry Commission Wales – *Stories from the Forest: Newsletter of the Forests in the Rural Community Project* (Forest Enterprise, 1st Edition – August 2002 and 2nd Edition – November 2002)
Hansard – *Afforestation, Towy Valley (Government Decision)* (HC Deb 31 January 1952 vol. 495, cc50-1W)
Jones, David J. V. – *Before Rebecca: Popular Protests in Wales 1793–1835* (Allen Lane, 1973)
— *Rebecca's Children: A Study of Rural Society, Crime, and Protest* (Oxford University Press, 1989)
Lovegrove, Roger – *The Kite's Tale: The Story of the Red Kite in Wales* (RSPB, 1990)
May, Celia A., Wetton, Jon H., Davis, Peter E., Brookfield, John F. Y. and Parkin, David T. – 'Single-Locus Profiling Reveals Loss of Variation in Inbred Populations of the Red Kite (*Milvus milvus*)' (*Proceedings of the Royal Society of London B*, 251, no. 1332, 1993)
—, Wetton, Jon H. and Parkin, David T. – 'Polymorphic Sex-Specific Sequences in Birds of Prey' (*Proceedings of the Royal Society of London B*, 253, no. 1338, 1993)
Owens, Nerys Elisa – 'The Shifting Governance of State Forestry in Britain: A Critical Investigation of the Transition from Productivism to Post-Productivism' (Thesis, School of City and Regional Planning, Cardiff University, 2009)
Spence, Barbara – *The Forestry Commission in Wales 1919–2013* (Forestry Commission Wales, 2013)
Thomas, Adrian L. R. – 'On the Tails of Birds' (Dissertation, Department of Ecology, Lund University, 1995)
Walpole-Bond, J. A. – *Bird Life in Wild Wales* (T. Fisher Unwin, 1903)
Walters Davies, P. and Davis, P. E. – 'The ecology and conservation of the Red Kite in Wales' (*British Birds*, 66, no. 5, 183–224, no. 6, 241–270, 1973)
Ward, Colin – *Cottars and Squatters: Housing's Hidden History* (Five Leaves, 2002)
Wildman, L., O'Toole, L. and Summers, R. W. – 'The diet and foraging behaviour of the Red Kite in northern Scotland' (*Scottish Birds*, 19, no. 3, 134–140, 1998)

XI Marsh Harrier

Axell, H. E. – 'Rare birds: the marsh harrier' (*Bird Life*, 7, no. 4, 98–101, 1971)

Burd, Fiona – *Erosion and vegetation change on the saltmarshes of Essex and north Kent between 1973 and 1988* (Nature Conservancy Council, 1992)
Buxton, Anthony – *Fisherman Naturalist* (Collins, 1946)
Chamberlain, Paul – *Hell Upon Water: Prisoners of War in Britain, 1793–1815* (The History Press, 2008)
Clarke, Roger – *The Marsh Harrier* (Hamlyn, 1995)
Daly, Augustus A. – *The History of the Isle of Sheppey* (Simpkin, Marshall, Hamilton, Kent & Co., 1894)
Gillham, E. H. and Homes, R. C. – *The Birds of the North Kent Marshes* (Collins, 1950)
Harrison, Jeffrey – *The Nesting Birds of Chetney Marsh* (Kent Ornithological Society, 1969)
— and Grant, Peter – *The Thames Transformed: London's River and its Waterfowl* (André Deutsch, 1976)
Hosking, Eric J. – 'Some Observations on the Marsh Harrier' (*British Birds*, 37, no. 1, 2–9, 1943)
Nature Conservancy Council – *Wildlife Conservation in the North Kent Marshes: A Report of a Working Party* (The Nature Conservancy, 1971)
Oliver, Peter – *Bird Watching on the North Kent Marshes* (Peter Oliver, 1991)
Riviere, B. B. – *A History of the Birds of Norfolk* (H. F. & G. Witherby, 1930)
Simmons, Robert E. – *Harriers of the World: Their Behaviour and Ecology* (Oxford University Press, 2000)
Underhill-Day, J. C. – 'Population and Breeding Biology of Marsh Harriers in Britain since 1900' (*Journal of Applied Ecology*, 21, no. 3, 773–787, 1984)
— 'The food of breeding Marsh Harriers (*Circus aeruginosus*) in East Anglia' (*Bird Study*, 32, no. 3, 199–206, 1985)
Weis, Henning – *Life of the Harrier in Denmark* (Wheldon & Wesley, 1923)

XII Honey Buzzard

Appleby, Ron – *European Honey-buzzards (Pernis apivorus) in North Yorkshire: A Breeding History on a Learning Curve: 1895–2010* (Hobby Publications, 2012)
Bijlsma, R. G. – 'Impact of severe food scarcity on breeding strategy and breeding success of Honey Buzzards (*Pernis apivorus*)' (*De Takkeling*, 6, no. 2, 107–118, 1998)
— 'Do Honey Buzzards (*Pernis apivorus*) produce pellets?' (*Limosa*, 72, no. 3, 99–103, 1999)
— 'Use and function of eyelids in European Honey-buzzards (*Pernis apivorus*)' (*De Takkeling*, 10, no. 2, 117–128, 2002)
— 'What is the predation risk for European Honey-buzzards (*Pernis apivorus*) in Dutch forests inhabited by food-stressed Northern Goshawks (*Accipiter gentilis*)?' (*De Takkeling*, 12, no. 3, 185–197, 2004)

— and Piersma, T. – 'Internal organs and gastrointestinal tract of European Honey-buzzards (*Pernis apivorus*) in comparison with non-insectivorous raptors' (*De Takkeling*, 10, no. 3, 214–224, 2002)
Buxton, Anthony – *Sporting Interludes at Geneva* (Country Life, 1932)
Clark, J. M. and Eyre, J. A. (eds) – *Birds of Hampshire* (Hampshire Ornithological Society, 1993)
Cobb, F. K. – 'Honey Buzzard at wasps' nest' (*British Birds*, 72, no. 2, 59–64, 1979)
Diermen, J. van, Manen, W. van and Baaij, E. – 'Habitat use, home range and behaviour of Honey Buzzards (*Pernis apivorus*) tracked on the Veluwe, central Netherlands, by GPS-loggers' (*De Takkeling*, 17, no. 2, 109–133, 2009)
Duff, Daniel G. – 'Has the plumage of juvenile Honey-buzzard evolved to mimic that of Common Buzzard?' (*British Birds*, 99, no. 3, 118–128, 2006)
Edlin, H. L. – *Trees, Woods and Man* (Collins, 1970)
Forestry Commission – *Tourism and Recreation Conflicts in the New Forest* (Forestry Commission Report, 2004)
Harrison, J. M. – 'Food of the Honey-Buzzard (*Pernis apivorus apivorus*)' (*Ibis*, 1, no. 4, 772–773, 1931)
Heuvelmans, Bernard – *On the Track of Unknown Animals* (Kegan Paul, 1995)
Holstein, Vagn – *Hvepsevaagen Pernis Apivorus Apivorus (L.)* (H. Hirschsprungs Forlag, 1944)
Irons, Anthony – *Breeding of the Honey Buzzard (Pernis apivorus) in Nottinghamshire* (Trent Valley Bird Watchers Nottinghamshire's Ornithological Society, 1980)
Koks, B., Straathof, A. and Bijlsma, R. G. – 'Common Wasps (*Vespula vulgaris*), insecticides and Honey Buzzards (*Pernis apivorus*)' (*De Takkeling*, 5, no. 3, 16–19, 1997)
Kostrzewa, Achim – *The Effect of Weather on Density and Reproduction Success in Honey Buzzards (Pernis apivorus)* (World Working Group on Birds of Prey and Owls, 1987)
Lascelles, Gerald – *Thirty-Five Years in the New Forest* (Edward Arnold, 1915)
Ogilvie, M. A. – 'European Honey-buzzards in the UK – correction to breeding totals' (*British Birds*, 96, no. 3, 145, 2003)
Pasmore, Anthony – *Verderers of the New Forest: A History of the New Forest 1877–1977* (Pioneer Publications, 1977)
Rackham, Oliver – *Woodlands* (Collins, 2010)
Roberts, S. J. and Lewis, J. M. S. – 'Observations of European Honey-buzzard breeding density in Britain' (*British Birds*, 96, no. 1, 37–39, 2003)
— and Coleman, M. – 'Some observations on the diet of European Honey-buzzards in Britain' (*British Birds*, 94, no. 9, 433–436, 2001)
—, Lewis, J. M. S. and Williams, I. T. – 'Breeding European Honey-Buzzards in Britain' (*British Birds*, 92, no. 7, 326–346, 1999)

Spradbery, J. Philip – *Wasps: An Account of the Biology and Natural History of Solitary and Social Wasps with Particular Reference to those of the British Isles* (Sidgwick & Jackson, 1973)
Trap-Lind, Ib – 'Observations on a Honey Buzzard digging out a wasps' nest' (*British Birds*, 55, no. 1, 36, 1962)
Tubbs, Colin R. – *The New Forest* (Collins, 1986)
White, Gilbert – *The Natural History of Selborne* (Letter XLIII) (Dent, 1974)
Wiseman, E. J. – 'Honey Buzzards in Southern England' (*British Birds*, 105, no. 1, 23–28, 2012)

XIII Hobby

Ashley, M. – 'On the Breeding Habits of the Hobby' (*British Birds*, 11, no. 9, 194–196, 1918)
Atherton, Peter F. – 'Barn Swallow giving specific alarm call for Hobby' (*British Birds*, 90, no. 11, 526, 1997)
Baker, J. A. – *The Hill of Summer* (Harper & Row, 1969)
Blomfield, R. M. – *Poole: Harbour, Heath and Islands* (Regency Press, 1984)
Chalmers, John – *Audubon in Edinburgh and his Scottish Associates* (NMS, 2003)
Chapman, Anthony – *The Hobby* (Arlequin Press, 1999)
Clarke, A., Prince, P. A. and Clarke, R. – 'The energy content of dragonflies (*Odonata*) in relation to predation by falcons' (*Bird Study*, 43, no. 3, 300–304, 1996)
Davis, Terence – *Arne: A Purbeck Parish in Peace and War* (Dorset Publishing Company, 2000)
Dobinson, Colin – *Fields of Deception: Britain's Bombing Decoys of World War II* (Methuen, 2000)
Fenech, Natalino – *Fatal Flight: The Maltese Obsession with Killing Birds* (Quiller Press, 1992)
Forty, George – *Frontline Dorset: A County at War 1939–45* (Dorset Books, 1994)
Fuller, R. J., Baker, J. K., Morgan, R. A., Scroggs, R. and Wright, M. – 'Breeding populations of the Hobby (*Falco Subbuteo*) on farmland in the southern Midlands of England' (*Ibis*, 127, no. 4, 510–516, 1985)
Green, George – *The Birds of Dorset* (Christopher Helm, 2004)
Insley, Hugh and Holland, Maurice G. – 'Hobbies feeding on bats, and notes on other prey' (*British Birds*, 68, no. 6, 242, 1975)
Nethersole-Thompson, Desmond – 'The Field Habits and Nesting of the Hobby' (*British Birds*, 25, no. 6, 142–150, 1931)
Trodd, Paul – 'Hobbies nesting on pylon' (*British Birds*, 86, no. 12, 625, 1993)
Walpole-Bond, John – *Field-Studies of Some Rarer British Birds* (Witherby & Co., 1914)
Webb, Nigel – *Heathlands* (Collins, 1986)

XIV Buzzard

Baring-Gould, S. – *A Book of Dartmoor* (Methuen, 1900)
Burnard, Robert – *Plundered Dartmoor* (Dartmoor Preservation Association, 1896)
Christensen, Steen, Nielsen, Bent Pors, Porter, R. F. and Willis, Ian – 'Flight identification of European raptors' (*British Birds*, 64, no. 6, 247–266, 1971)
Crossing, William – *A Hundred Years on Dartmoor* (The Western Morning News, 1901)
— *The Teign: From Moor to Sea* (Quay Publications, 1986)
Dare, Peter – *The Life of Buzzards* (Whittles, 2015)
English Nature and Dartmoor National Park Authority – *The Nature of Dartmoor: A Biodiversity Profile* (English Nature, 2001)
Fryer, G. – 'Aggressive behaviour by buzzards at nest' (*British Birds*, 67, no. 6, 238–239, 1974)
— 'Buzzard and crows at Magpie nest' (*British Birds*, 79, no. 1, 40–41, 1986)
Fryer, Geoffrey – 'Notes on the breeding biology of the Buzzard' (*British Birds*, 79, no. 1, 18–28, 1986)
Harvey, L. A. and St Leger-Gordon, D. – *Dartmoor* (Fontana, 1974)
Hemery, Eric – *High Dartmoor: Land and People* (Robert Hale, 1983)
Martin, E. W. – *Dartmoor* (Robert Hale, 1958)
Moore, N. W. – 'The Past and Present Status of the Buzzard in the British Isles' (*British Birds*, 50, no. 5, 173–197, 1957)
Picozzi, N. and Weir, D. – 'Breeding biology of the Buzzard in Speyside' (*British Birds*, 67, no. 5, 199–210, 1974)
— 'Dispersion of Buzzards in Speyside' (*British Birds*, 76, no. 2, 66–78, 1983)
Prytherch, Robin – 'The breeding biology of the common buzzard' (*British Birds*, 106, no. 5, 264–279, 2013)
Ryves, B. H. – *Bird Life in Cornwall* (Collins, 1948)
Tubbs, Colin R. – *The Buzzard* (David & Charles, 1974)
Tubbs, C. R. – 'Population study of Buzzards in the New Forest during 1962–66' (*British Birds*, 60, no. 10, 381–395, 1967)
Wareing Ormerod, G. – 'Notes on the Geology of the Valleys of the Upper Parts of the River Teign and its Feeders' (*Proceedings of the Geological Society*, November 1867)

XV Sparrowhawk

Baker, J. A. – *The Hill of Summer* (Harper & Row, 1969)
Barlow, Nora (ed.) – *The Autobiography of Charles Darwin* (Collins, 1958)
Deane, Ruthven – 'Unpublished Letters of William MacGillivray to John James Audubon' (*The Auk*, 18, 239–249, 1901)
Newton, Alfred – *A Dictionary of Birds* (Adam & Charles Black, 1896)
Newton, Ian – *The Sparrowhawk* (T. & A. D. Poyser, 1986)

Newton, I., Bell, A. A. and Wyllie, I. – 'Mortality of Sparrowhawks and Kestrels' (*British Birds*, 75, no. 5, 195–204, 1982)

Newton, I., Dale, L. and Rothery, P. – 'Apparent lack of impact of Sparrowhawks on the breeding densities of some woodland songbirds' (*Bird Study*, 44, no. 2, 129–135, 1997)

Nicholas, W. W. – *A Sparrow-hawk's Eyrie* (A. Brown & Sons, 1937)

Owen, J. H. – 'Some Breeding-Habits of the Sparrow-Hawk' (*British Birds*, 15, no. 4, 74–77, no. 11, 256–263, 1921–22)

— 'The Feeding-Habits of the Sparrow-Hawk' (*British Birds*, 25, no. 6, 151–155, 1931)

— 'The Hunting of the Sparrow-Hawk' (*British Birds*, 25, no. 9, 238–243, 1932)

— 'The Food of the Sparrow-Hawk' (*British Birds*, 26, no. 2, 34–40, 1932)

Perrins, C. M. and Geer, T. A. – 'The Effect of Sparrowhawks on Tit Populations' (*Ardea*, 68, 133–142, 1980)

Pounds, Hubert E. – 'Notes on the Flight of the Sparrow-Hawk' (*British Birds*, 30, no. 6, 183–189, 1936)

Tinbergen, Lukas – 'De Sperwer als Roofvijand van Zangvogels/The Sparrowhawk as a Predator of Passerine Birds' (*Ardea*, 34, 1–213, 1946)

Yapp, W. B. – *Birds and Woods* (Oxford University Press, 1962)

General

Aberdeen Free Press – 'In Memoriam. William MacGillivray, M.A., LL.D., Professor of Natural History and Lecturer on Botany, in Marischal College and University, Aberdeen; 1841–1852' (*Aberdeen Free Press*, 21 November 1900)

Audubon, John James – *Ornithological Biography*, volumes I–V (Adam & Charles Black, 1831–39)

Balmer, D. E, Gillings, S., Caffrey, B. J., Swann, R. L., Downie, I. S. and Fuller, R. J. – *Bird Atlas 2007–11: The breeding and wintering birds of Britain and Ireland* (BTO Books, 2013)

Bannerman, David A. – *The Birds of the British Isles*, volume V (Oliver & Boyd, 1956)

Baxter, Evelyn V. and Rintoul, Leonora J. – *The Birds of Scotland: Their History, Distribution, and Migration* (Oliver and Boyd, 1953)

Bijleveld, Maarten – *Birds of Prey in Europe* (Macmillan, 1974)

Bircham, Peter – *A History of Ornithology* (Collins, 2007)

Birkhead, Tim – *Bird Sense: What It's Like to Be a Bird* (Bloomsbury, 2012)

Broun, Maurice – *Hawks Aloft: The Story of Hawk Mountain* (Dodd, Mead, 1949)

Brown, Leslie – *British Birds of Prey* (Collins, 1976)

— *Birds of Prey: Their Biology and Ecology* (Hamlyn, 1976)

Cade, Tom J. – *Falcons of the World* (Collins, 1982)

Campbell, Bruce and Ferguson-Lees, James – *A Field Guide to Birds' Nests* (Constable, 1972)

Chalmers, John – *Audubon in Edinburgh and his Scottish Associates* (NMS, 2003)
Cobham, David and Pearson, Bruce – *A Sparrowhawk's Lament: How British Breeding Birds of Prey are Faring* (Princeton University Press, 2014)
Cocker, Mark and Mabey, Richard – *Birds Britannica* (Chatto & Windus, 2005)
Corning, Howard (ed.) – *Letters of John James Audubon 1826–1840* (The Club of Odd Volumes, 1930)
Craighead, John J. and Craighead, Frank C. – *Hawks, Owls and Wildlife* (Dover, 1969)
Deane, Ruthven – 'Unpublished Letters of William MacGillivray to John James Audubon' (*The Auk*, 18, 239–249, 1901)
Fairbrother, Nan – *New Lives, New Landscapes* (Penguin, 1972)
Ferguson-Lees, James and Christie, David A. – *Raptors of the World* (Christopher Helm, 2001)
Frederick II, Emperor – *The Art of Falconry being the De Arte Venandi Cum Avibus* (Oxford University Press, 1955)
Gordon, Seton – *Hill Birds of Scotland* (Edward Arnold, 1915)
— *Wild Birds in Britain* (Batsford, 1938)
— *In Search of Northern Birds* (Eyre & Spottiswoode, 1941)
Greenoak, Francesca – *British Birds: Their Folklore, Names and Literature* (Christopher Helm, 1997)
Hardey, Jon, Crick, Humphrey Q. P., Wernham, Chris V., Riley, Helen T., Etheridge, Brian and Thompson, Des B. A. – *Raptors: A Field Guide to Survey and Monitoring* (Stationery Office, 2006)
Harkness, Roger and Murdoch, Colin – *Birds of Prey in the Field: A Guide to the British and European Species* (Witherby, 1971)
Harrison, J. C. – *Bird Portraits* (Country Life, 1949)
Hart-Davis, Duff – *Audubon's Elephant: The Story of John James Audubon's Epic Struggle to Publish The Birds of America* (Phoenix, 2004)
Holloway, Simon – *The Historical Atlas of Breeding Birds in Britain and Ireland: 1875–1900* (T. & A. D. Poyser, 1996)
Jonsson, Lars – *Birds of Europe with North Africa and the Middle East* (Christopher Helm, 1992)
Lensink, Rob – 'Range expansion of raptors in Britain and the Netherlands since the 1960s: testing an individual-based diffusion model' (*Journal of Animal Ecology*, 66, no. 6, 811–826, 1997)
MacGillivray, William – 'Journal of a year's residence and travels in the Hebrides by William MacGillivray from 3rd August 1817 to 13th August 1818 Vol. I' (Manuscript, University of Aberdeen Special Collections, 1817–1818)
— 'Notes taken in the course of a journey from Aberdeen to London by Braemar, Fortwilliam, Inveraray, Glasgow, Ayr, Dumfries, Carlisle, Keswick, Kendal, Manchester, Derby and Northampton in 1819' (Manuscript, University of Aberdeen Special Collections, 1819)

— (trans.) – *Elements of Botany and Vegetable Physiology by A. Richard* (William Blackwood, 1831)
— 'Report on Visit to Museums' (Manuscript, University of Aberdeen Special Collections, 1833)
— *Lives of the Eminent Zoologists, From Aristotle to Linnaeus* (Oliver & Boyd, 1834)
— (ed.) – *The Edinburgh Journal of Natural History, and of the Physical Sciences* (The Edinburgh Printing Company, 1835)
— *Descriptions of the Rapacious Birds of Great Britain* (Maclachlan & Stewart, 1836)
— 'Autographic Notes Chiefly on British Birds' (Manuscript, University of Aberdeen Special Collections, 1836–1840)
— *A History of British Birds, Indigenous and Migratory*, volumes I–V (Scott, Webster, and Geary, 1837–52)
— *A History of British Quadrupeds* (W. H. Lizars, 1838)
— *The Travels and Researches of Alexander Von Humboldt* (Harper & Brothers, 1839)
— *A Manual of British Ornithology: Being a Short Description of the Birds of Great Britain and Ireland, Including the Essential Characters of the Species, Genera, Families, and Orders* (Scott, Webster, and Geary, 1840)
— *A Manual of Botany: Comprising Vegetable Anatomy and Physiology, or the Structure and Functions of Plants* (Scott, Webster, and Geary, 1840)
— *A Manual of Geology* (Scott, Webster, and Geary, 1840)
— 'Testimonials in favour of William MacGillivray, A.M.' (University of Aberdeen Special Collections, 1841)
— *A History of the Molluscous Animals of the Counties of Aberdeen, Kincardine, and Banff* (Cunningham & Mortimer, 1843)
— (ed.) – *A Systematic Arrangement of British Plants by W. Withering* (Adam Scott, 1845)
— (ed.) – *The Conchologist's Text-Book by T. Brown* (A. Fullerton, 1845)
— *Manual of British Birds: Including the Essential Characters of the Orders, Families, Genera, and Species* (Adam Scott, 1846)
— *The Natural History of Dee Side and Braemar* (Private publication, 1855)
— *A Hebridean Naturalist's Journal 1817–1818* (Acair, 1996)
— *A Walk to London* (Acair, 1998)
MacGillivray, William and Thomson, J. Arthur – *Life of William MacGillivray* (John Murray, 1910)
MacGillivray, William – *A Memorial Tribute to William MacGillivray* (Edinburgh, 1901)
Mead, Chris – *The State of the Nations' Birds* (Whittet Books, 2000)
Meinertzhagen, R. – *Pirates and Predators: The Piratical and Predatory Habits of Birds* (Oliver & Boyd, 1959)
Meyburg, Bernd-Ulrich and Chancellor, R. D. (eds) – *Raptors in the Modern World* (World Working Group on Birds of Prey and Owls, 1987)

Mullarney, Killian, Svensson, Lars, Zetterström, Dan, Grant, Peter J. – *Collins Bird Guide* (HarperCollins, 1999)
Mynott, Jeremy – *Birdscapes: Birds in Our Imagination and Experience* (Princeton University Press, 2009)
Newton, Alfred – *Ootheca Wolleyana: An Illustrated Catalogue of the Collection of Birds' Eggs Formed by the Late John Wolley, Part I. Accipiters* (John Van Voorst, 1864)
Newton, Ian – *Population Ecology of Raptors* (T. & A. D. Poyser, 1979)
— 'Raptors in Britain – A review of the last 150 years' (*BTO News*, 131, 6–7, 1984)
Nicholson, E. M. – *Birds in England: An Account of the State of our Bird-Life and A Criticism of Bird Protection* (Chapman & Hall, 1926)
Porter, R. F., Willis, Ian, Christensen, Steen, Nielsen, Bent Pors – *Flight Identification of European Raptors* (T. & A. D. Poyser, 1981)
Proctor, Noble S. and Lynch, Patrick J. – *Manual of Ornithology: Avian Structure & Function* (Yale University Press, 1993)
Ralph, Robert – *William MacGillivray* (HMSO, 1993)
— *William MacGillivray: Creatures of Air, Land and Sea* (Merrell Holberton, 1999)
Richmond, Kenneth W. – *British Birds of Prey* (Lutterworth, 1959)
Ritchie, James – *The Influence of Man on Animal Life in Scotland* (Cambridge University Press, 1920)
RSPB – *Birds of Prey in the UK: Back from the Brink* (RSPB, 1997)
— *Birds of Prey in the UK: On a Wing and a Prayer* (RSPB, 2007)
— *Birdcrime: Offences Against Wild Bird Legislation in 2012* (RSPB, 2012)
Self, Andrew – *The Birds of London* (Bloomsbury, 2014)
Snow, D. W. and Perrins, C. M. – *The Birds of the Western Palearctic*, Volume 1, *Non-Passerines* (Oxford University Press, 1998)
Thompson, D. B. A., Redpath, S. M., Fielding, A. H., Marquiss, M., Galbraith, C. A. (eds) – *Birds of Prey in a Changing Environment* (Scottish Natural Heritage, 2003)
Thompson, Des, Riley, Helen and Etheridge, Brian – *Scotland's Birds of Prey* (Lomond, 2010)
Videler, John J. – *Avian Flight* (Oxford University Press, 2005)
Walpole-Bond, John – *Field-Studies of Some Rarer British Birds* (Witherby & Co., 1914)
Watson, Donald – *Birds of Moor and Mountain* (Scottish Academic Press, 1972)
Whitlock, Ralph – *Rare and Extinct Birds of Britain* (Phoenix House, 1953)
Yarrell, William – *A History of British Birds*, volumes I–III (John Van Voorst, 1845)
Yosef, R., Miller, M. L. and Pepler, D. (eds) – *Raptors in the New Millennium* (International Birding and Research Center, 2002)
Zuberogoitia, Iñigo and Martínez, José Enrique (eds) – *Ecology and Conservation of European Forest-Dwelling Raptors* (Diputación Foral de Bizkaia, 2011)

Acknowledgements

I am grateful to The Royal Society of Literature and The Jerwood Charitable Foundation for an RSL Jerwood Award for Non-Fiction in 2011, and to The Society of Authors for an Authors' Foundation Roger Deakin Award in 2011.

Thank you to my agent Jessica Woollard at The Marsh Agency; Nicholas Pearson, my editor at Fourth Estate, and Kate Tolley at Fourth Estate; Andrew McNeillie for his support and encouragement and for publishing work in progress from this book in *Archipelago* magazine; Robert Macfarlane and Robin Harvie for their encouragement; Antony Harwood for his patient support.

Thank you also to Sophie Wilcox at the Alexander Library of Ornithology in Oxford; Jim McGregor at the University of Aberdeen Herbarium; the staff at the Special Collections Library, University of Aberdeen; the staff at the Natural History Museum Library, London; Alyson Tyler at CyMAL Museums Archives and Libraries Wales; George Deane in Bolton; Bob Image and Jim Scott in East Anglia; Eric Meek on Orkney; David Pearce in Gloucestershire; Norrie Russell in Sutherland.

Above all, my thanks to Nicki.